I0491166

Sheep Breeds and Sheep Management
A Livestock Handbook on Sheep

by John Wrightson

with an introduction by Jackson Chambers

This work contains material that was originally published in 1898.

This publication is within the Public Domain.

This edition is reprinted for educational purposes
and in accordance with all applicable Federal Laws.

Introduction Copyright 2018 by Jackson Chambers

Self Reliance Books

Get more historic titles on animal and stock breeding, gardening and old
fashioned skills by visiting us at:

http://selfreliancebooks.blogspot.com/

Introduction

I am pleased to present yet another practical title on breeding and raising livestock.

The work is in the Public Domain and is re-printed here in accordance with Federal Laws.

As with all reprinted books of this age that are intended to perfectly reproduce the original edition, considerable pains and effort had to be undertaken to correct fading and sometimes outright damage to existing proofs of this title. At times, this task is quite monumental, requiring an almost total "rebuilding" of some pages from digital proofs of multiple copies. Despite this, imperfections still sometimes exist in the final proof and may detract from the visual appearance of the text.

I hope you enjoy reading this book as much as I enjoyed making it available to readers again.

Jackson Chambers

CONTENTS.

—o—

ILLUSTRATIONS.

—o—

LINCOLN RAMS.

SHROPSHIRE RAM.

DARTMOOR SHEEP.

HERDWICK RAM.

Cheviot Ram.

BLACK-FACED RAM.

DORSET HORN RAM.

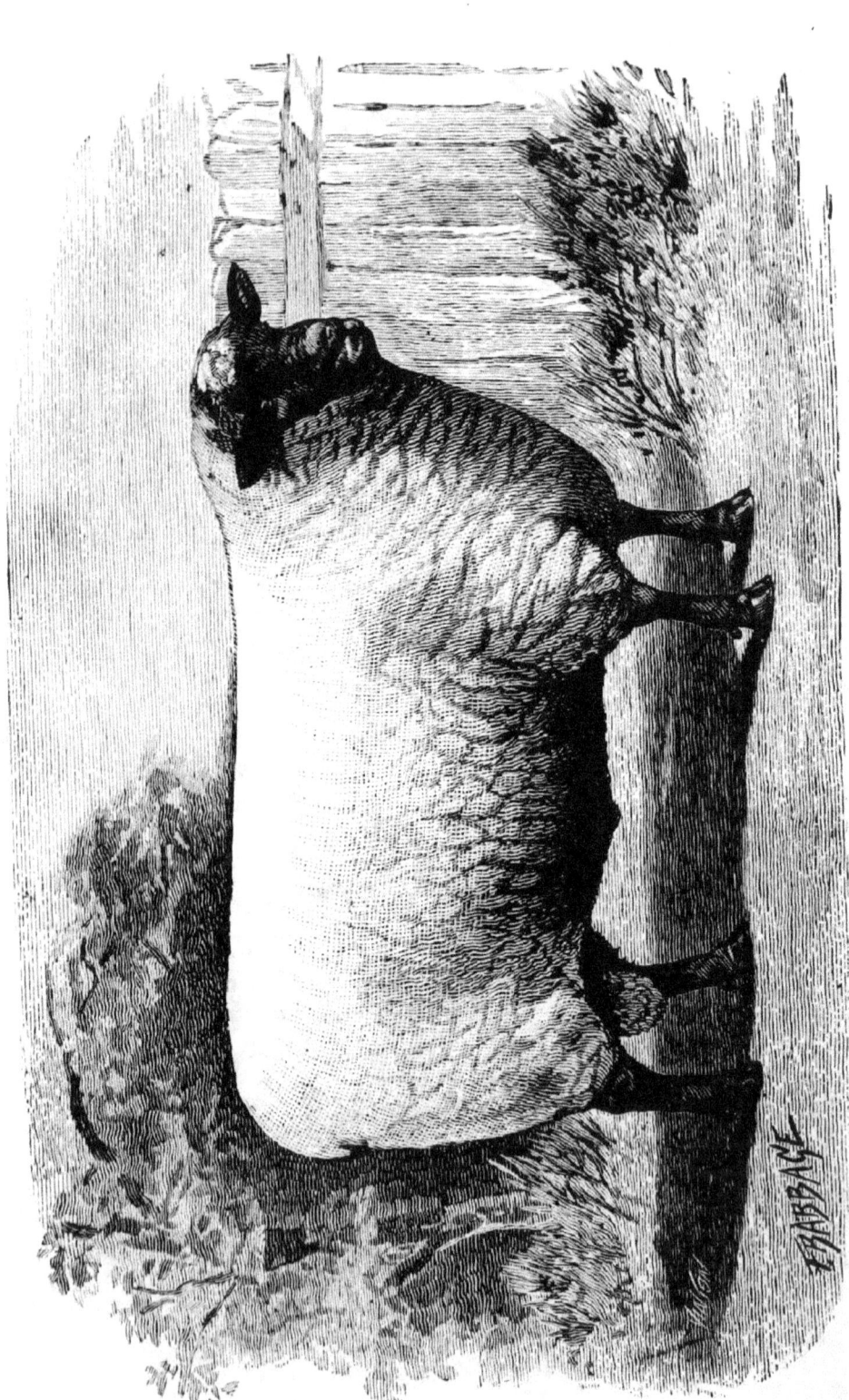

SUFFOLK RAM.

SHROPSHIRE RAM.

SHROPSHIRE RAMS.

OXFORD DOWN RAM.

HAMPSHIRE DOWN FAT WETHERS.

HAMPSHIRE DOWN RAM.

SOUTHDOWN RAM.

SOUTHDOWN EWES.

SOUTHDOWN RAM.

WENSLEYDALE RAM.

DEVON LONGWOOL RAM.

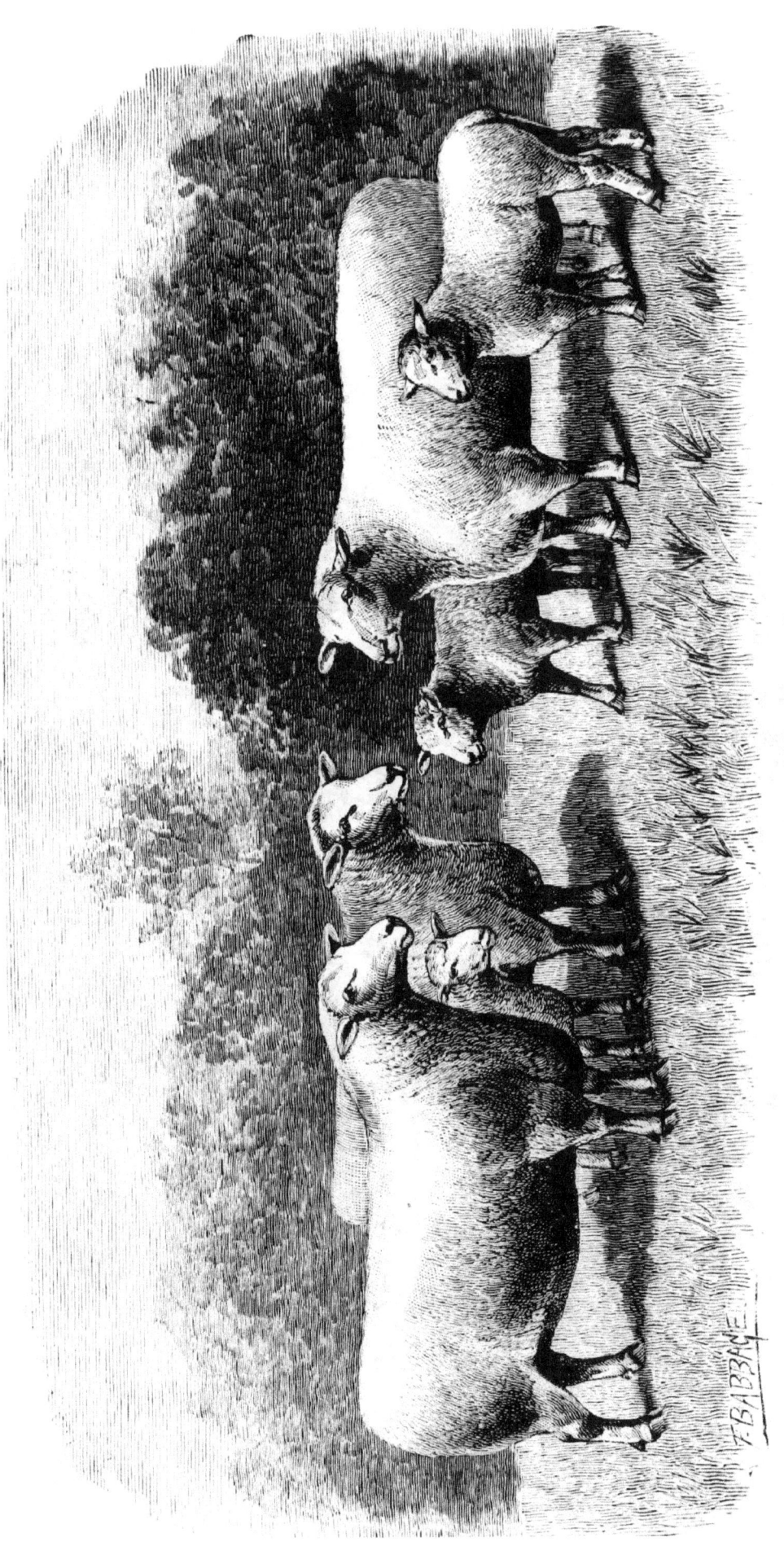

Kent or Romney Marsh Ewes and Lambs.

COTSWOLD RAM.

BORDER LEICESTER EWE.

LINCOLN RAM.

LEICESTER RAM.

MERINO RAM.

SHEEP.

BREEDS AND MANAGEMENT.

PROLOGUE.

THERE is genuine poetry in pastoral life which it would be sad to lose. Nevertheless, agricultural science and literature are between them rapidly taking the romance out of it. Perhaps we should add, hard times, and the vital importance of making things "pay." It is a pity to lose the faculty of discerning the beauty of "the dewy eve and rising moon," or listening as "the amorous thrush concludes his song"; or only to think of the price of mutton and of wool, or of lambs as fore and hind quarters.

> The haunt o' Spring's the primrose brae,
> The Summer's joy the flocks to follow;
> How cheerie through the shortening day
> Is Autumn in her weeds o' yallow !

The sweetness of pastoral life is going. It is disappearing under the influence of commercial enterprise, the spread of science, and the difficulties of competition. We ourselves are victims to utilitarianism, and must plead guilty to sharing in the universal want of sentiment, even when "birds rejoice in leafy bowers, and bees hum round the breathing flowers;" or when "within yon milk-white hawthorn bush, among her nestlings sits the thrush." One is sometimes inclined to wonder if steam power and chemical manures, pedigree stock and iron fencing, weigh-bridges and milk registers, will ever compensate us for the loss of the fresh and simple country

2 ᴵ

life of our forefathers. It is useless to repine, and perhaps the best thing we can do is to cherish those pleasurable feelings with which we may still view the flock spread o'er the down, or listen to the varied tones of the sheep bell; and to cultivate more of personal interest and affection for our domesticated creatures. There is no doubt that the humble dairyman, the carter, and the shepherd, obtain more enjoyment from watching and tending their charges than do their masters; and the pleasures of farming might be greatly enhanced by devoting more personal attention to live stock, and studying their habits. Love of animals may be cultivated, and with it comes an interest in the creatures which surround us.

A few beans in the pocket for the horses, a few pets among the sheep, or a handful of corn thrown to the poultry, soon beget love on both sides, and is not thrown away, even where profit is concerned. Interest and kindness on the part of the master insure and promote similar interest and kindness on the part of the servants, and help to form a good judgment as to the merits of the animals. A judge of dogs and of horses must be really fond of these animals, and unless a man is fond of cattle, sheep, and even pigs, he can hardly hope to excel, either as a manager or as a breeder of stock. Individual characters only become apparent after long observation, and hence we recommend the cultivation of kindly feeling and real interest in the welfare of individual animals, as one way at least of preventing the merely commercial feeling from absorbing us too exclusively.

In commencing a book on sheep, we urge the importance of real interest in the animals themselves. The shepherd, if worthy of his position, knows his flock individually; and it is absolutely necessary that he should so know his sheep, if progress is to be made. This intimate knowledge can scarcely be expected of the master, but he should at least sympathise with his shepherd, and maintain such an interest in his sheep as to encourage him in his work. Frequent visits to the sheepfolds, and talks with the shepherd as to the well-being of his charge, are in themselves interesting and also instructive to both parties. The knowledge possessed by the shepherd is of a different sort to that of the master, but both are equally useful, and the exchange of ideas is mutually beneficial.

CHAPTER I.

THE EFFECTS OF DOMESTICATION.

THAT our domesticated animals are more or less related tc certain wild forms is evident; but it is also true that in many cases we are unable to trace any close resemblance between a highly-bred domesticated specimen and its wild relations. It is not our intention at present to dilate upon this topic, except to indicate the fact that similar variations are to be found wherever a wild form has been domesticated. Nay more, a similar diversity of type, brought about by cultivation, is to be noted in the products of our gardens, whether among ornamental flowers and shrubs, useful vegetables, or fruits. Varieties are endless, and in many cases can be traced to changes in one normal or aboriginal type. Roses, dahlias, pansies, and begonias among flowers; cabbages, broccolis, cauliflowers, and potatoes among familiar vegetables; apples, pears, and peaches among fruits; each appears to have been originally descended and improved from more or less unpromising prototypes. Taking a similar general view of our live stock and domestic pets, we see many different descriptions of dogs, cats, and pigeons, or even of mice or rats, when bred in confinement. Wherever, indeed, the hand of man has been laid upon an animal or a vegetable we witness a strange metamorphosis, not only in the form of varieties already in existence, but in the easy formation of new ones. This liability to vary when animals are domesticated, or plants cultivated, is an accepted fact by all naturalists and all breeders of fancy stock, and in its light it is not difficult to imagine that all our breeds of sheep may have been descended from one wild form. That this has actually been the case cannot be asserted, and in fact is scarcely likely to be true. Wild species of all animals vary with the conditions of climate and food under which they live. The original native sheep of England, if such there was, would probably be slightly different from the wild sheep of the Continent of Europe or of Asia, or even from the original sheep of northern and of southern Europe.

Whether our sheep were originally brought from Asia in the train of conquering armies, as was no doubt the case with the ancestors of many of our domesticated animals, may be considered an open question, excepting that we can hardly imagine that they were absent from those vast migrations of the human family. They, however, existed in Europe in the Stone period, although in less number than goats. The probability is that domesticated sheep originated through the domestication of several races in many parts of the world, the peculiarities and valuable properties of each having been developed by selection, until a more or less perfect type was obtained.

It is one of the peculiarities induced by domestication that animals, although descended from different stocks, which in the wild state would not have mated together, breed freely, and produce fertile offspring. Such is the case between the *sus scrofa*, or European pig, and the specifically distinct *sus indica*, or Chinese hog. Also between the buffalo and bison, and ordinary cattle. It has also been shown to have been the case between the different species of oxen from which our European cattle are now usually supposed to be descended.

The crossing of species of sheep originally distinct has no doubt still further increased the number of our recognised breeds.

Lastly, there is the continued breeding of animals by men who have had a particular object and type of animal in view. The cattle possessed by the different tribes of Caffres of

Southern Africa have been observed to differ from one another, as is shown by the facility with which the natives discriminate them. In our own country the Booth and Bates tribes among Shorthorns, the varieties of Devon cattle, and (although now merged in a common type) the Wilts and Hants sheep, are evidence that when animals of even the same race are bred for a long period in different hands, the original type alters considerably.

We have now indicated the means by which the numerous breeds, not only of sheep but of other domestic animals, have been produced, and we turn to the important topic of the

VARIOUS BREEDS OF SHEEP.

Before describing the breeds *seriatim*, it will be useful to take a general survey of the fleecy inhabitants of these islands. Commencing with the far north, we find the hills of the Highlands and islands of Scotland tenanted by Black-faced sheep, where these have not given way to the more majestic denizens of deer forests. The Black-faced sheep flourish wherever heather grows, and their picturesque appearance agrees well with the land of the mountain and the flood. The lower slopes of the hills are tenanted by Cheviots, and these two races share the greater part of Scotland between them. In the Lowlands we find Cheviots, Border Leicesters, and crosses of Cheviot and Leicester occupying most of the arable farms. In the arable portions of Cumberland and Northumberland a similar description of sheep is kept, the ewe flocks being generally half or three parts Leicester put to Leicester tups, the remaining blood being Cheviot. On the hills of Cumberland and Westmoreland the hardy little Herdwick is the favourite; while on the extensive moors stretching from the Cheviots to Allen Heads, and onwards through South Northumberland, Durham, Yorkshire, and to the Peak district of Derbyshire, the Black-faced breed again predominates. The hills of West Yorkshire and East Lancashire are tenanted by the Limestone or Crag sheep, while the mossy valleys carry the damp-enduring Lonks, with their black and white faces and superior wool. The poor arable lands of Durham and North Yorkshire are stocked with "mules," the progeny of Black-faced ewes and Leicester rams. Throughout East Yorkshire, Notts, Huntingdon, Rutland, Northampton, South Derbyshire, Leicestershire, and parts of Lincolnshire, the Leicester sheep is still the favourite. The Lincoln sheep is found in the greatest perfection in the southern portions of their county. There are many also in the northern part of Cambridgeshire. In the Western Midlands, including Shropshire, Staffordshire, Warwickshire, parts of Cheshire, and Worcestershire, the famous Shropshire sheep has achieved a prominent position; while in portions of Herefordshire, Monmouth, Gloucestershire, and parts of Oxfordshire and Somersetshire, the Cotswold sheep reigns. In Oxfordshire, Berks, Bedford, Bucks, and Herts, the improved Oxford Down is the characteristic sheep. In Wiltshire and Hampshire, the Hampshire Down predominates. In Sussex, Kent, and Surrey, we find the natural home of the Southdown, but in the southern parts of Kent, the rich pastures of Romney Marsh and the adjoining uplands support a long-woolled race of their own. In Suffolk and Essex, the Suffolk Down sheep has recently reasserted its claim to be recognised as a separate and excellent race. On the extreme south coast of Hampshire, and in the county of Dorset, the white-faced Dorsets or "horns" prevail, while in Devon the Devon Long Wool breed is much esteemed. On the hills and moors of the same county are to be found the hardy Exmoor and Okehampton forest breeds. In the Principality many of the already named races have found a footing on the lowlands, while the mountains are inhabited by Scotch Black-faces and Welsh sheep of small size, but boasting exceptionally good mutton.

The above sketch of localities of the various breeds must not be understood as more than general. In every county are to be found flocks of Shropshires, Oxfords, Hampshires, Cotswolds, or Southdowns, but the general stocking of farms will be found to be upon the lines indicated.

DISTRIBUTION OF SHEEP.

Taking the United Kingdom as a whole we find that the general distribution of sheep is represented at about 600 sheep per 1,000 acres of agricultural land, associated with 214 cattle of all ages, 40 horses, and 80 pigs upon the same area.

The relative sheep populations of the various portions of the United Kingdom are shown as follows :—

England maintains	632 sheep per 1,000 acres.				
Scotland	„	1,380	„	„	„
Wales	„	964	„	„	„
Ireland	„	241	„	„	„

We see, then, that Caledonia is the land for sheep and shepherds, and of the delights of arcadian and pastoral life. Next to her comes "gallant little Wales," while England is evidently more conspicuously divided between sheep, cattle and corn-growing.

The following are the principal English sheep-breeding counties according to the most recent agricultural returns :—

County.				Total area.			Total Number of Sheep.
Lincoln	...	...	...	1,767,962	...	...	1,206,684
Northumberland	...		...	1,290,312	...	...	914,973
Kent	...	...	...	1,004,984	...	...	896,566
Devonshire		...	...	1,655,161	...	...	835,441
Wilts	...	...	...	859,303	...	...	626,132
Dorset	...	...	...	627,265	...	...	426,254

Although Scotland possesses so many sheep, the difference in fertility between that country and of England is readily seen upon examining statistics as to the number of sheep maintained. Thus we find, on glancing at the most recent agricultural statistics of Great Britain, the following figures representing the number of sheep in various Scotch counties:—

County.				Total Area.			Total Number of Sheep.
Aberdeen	...	...	...	1,258,510	...	...	165,023
Argyll	...	...	...	2,092,458	...	...	972,737
Inverness	...	...	...	2,708,237	...	...	616,285
Perth...	...	...	...	1,656,082	...	...	694,647
Sutherland	...	...	...	1,347,033	...	...	196,871

If we contrast Northumberland with Aberdeenshire, Lincoln with Perth, Kent with Sutherlandshire, we shall at once see the immense difference in productive power between the fertile English soil and the barren uplands of the North of Scotland.

Still, as already shown, Scotland is far ahead of England as a sheep-breeding country, and this conclusion is not falsified by the few instances just given in which the mountains of Aberdeen, Perth, and Sutherland are contrasted with the richer soils of Northumberland, Lincoln, and Kent.

CHAPTER II.

LONG AND FINE WOOLLED SHEEP.

ALL sheep may be spoken of as long, middle, or short-woolled. The classification has more merit than at first appears, because with the length of the fleece are associated other peculiarities which help to form distinct classes. Thus in this country the short-woolled sheep include the Down breeds, with their brown faces, high quality of mutton, and active natures. The long-woolled breeds, on the other hand, are white-faced, somewhat coarse in flesh, and more indolent in their habits.

MERINO SHEEP.

Although we properly divide our flocks into long and short woolled races, we are in England destitute of a truly short or fine woolled breed. If we want to see short and fine wool to perfection we must seek it among the Merino flocks of Southern Europe, or our own colonies of Australia and New Zealand. The early home of Merino sheep appears to have been Spain,* from whence they were imported into France, England, Hungary, and Germany. The moist climate of Britain was, however, unfavourable for the growth of the

---

* Professor Low says that the sheep of Spain were introduced at various periods : first, from Asia by the early Phœnician Colonies ; second, from Africa by the Carthaginians ; third, from Italy by the Romans ; and fourth, from Africa, by the Moors, during nearly eight centuries of occupation. The most important of these latter races is the Merino, now the most esteemed and widely diffused of all the fine woolled breeds of Europe.

finest wools, and hence the Merino has never been successfully propagated with us. It was soon found that the dry
climates of Australasia were eminently adapted for the production of fine wool, and hence the vast flocks of Australia
and New Zealand are composed of Merino and Merino crosses,
and the wool which reaches us from these colonies is all of
the finest description.

The Merino is eminently a *wool* sheep, as distinct from a
mutton sheep. For feeding purposes it is inferior ; and as
the production of the finest wool is best attained by feeding
on rye straw and a spare diet, it is evident that where fine
wool is the object, heavy feeding is at a discount. Since the
frozen mutton trade has attained its recent proportions, increased attention has been given to the carcases of sheep in
the leading sheep-farming colonies, and the Merino has been
largely crossed with Lincolns, Shropshires, Leicesters, and
Southdowns, with the effect of greatly improving the value of
the animals, although at a sacrifice of the quality of the wool.
The bulk of the wool, however, imported from Australia is
decidedly of the Merino type.

English farmers have but little idea of the importance
attached to the quality of wool by Merino breeders. The
dividing of a flock according to the varying qualities of the
wool is not attempted by the breeder of Austro-Hungary,
but is committed to competent experts, who visit the stations
at the proper time of year (April), and carefully examine and
place the animals in classes. Rams are selected chiefly for
their wool, and the care and amount of discussion over
this product would surprise some of our sheep breeders.

Merino sheep have been divided into Rambouillet, Negretti
and Electoral. Respecting this classification, the " Weiner
Landwirthschaftlicher Zeitung," when reporting upon the
Vienna Exhibition of 1873, said: " We found the names were
often wrongly applied. We could only discern two principles
or ideas—the production of clothing wool and combing wool.
The difference between Electoral and Negretti which existed
fifty years ago is lost."

The Rambouillet Merino is of French origin. The character of the wool is fine, but much longer than the Negretti type, and the weight yielded per head is much heavier. The skin of the Rambouillet is also free from those characteristic wrinkles which are a feature in the Spanish Merino.

The selection of ewes and rams for the proper preservation of the quality of the wool has been already alluded to. All extensive flocks are divided into sections differing from each other in the quality of their fleeces. A large flock is divided as follows:—(1) Prima, (2) super-prima, (3) elector, (4) super-elector, (5) super-super-elector, and in some cases a still higher grade entitled "super-super-super-elector." The term *secunda* is also employed to indicate a grade inferior to the prima, making in all seven degrees of fineness. The skill of the classifier is shown in his recognition of the following points:—Strength, or what is usually spoken of as *kraft*, indicated by the amount of grease in the wool. The fat exists in three forms, soft or liquid, middle fat, and broken stiff fat, and in each case the colour may be yellow or white. 2. The wool should be equally fine over the whole body. A coarser quality may be expected on the rump and tops of the shoulders, and a weaker quality under the belly. 3. Curl, or the minute crinkles or waves in the wool fibre which give elasticity to the fleece. Both a long hair-like fibre, as at B (see page 13), and a too abrupt curl in which the waves are reduplicated or curl back upon themselves, as at C, are to be avoided. The best form of curl is a uniform crinkle, as at A. 4. Thickness of wool upon the skin. A Merino will carry 40,000 to 48,000 wool fibres upon a square inch of skin. The wool-bearing surface is also considerably increased in the Spanish Merinos by the folds of skin on the neck and over the rump. 5. The closure of the wool at the surface is very important, as an open fleece lets in dirt and foreign bodies, whereas a well-closed fleece presents the appearance of a continuous surface. 6. The wool must grow over the entire body down to the claws, and wool fibres are often seen

growing out of the bottom of the hoofs. The head, down to
the nose and over the ears, must all be completely covered
with wool. 7. The length of the wool varies from one to four
finger breadths, but is longer in the Rambouillet type.
8. The last point is the stature and carcase points of the
sheep. Under this heading the sheep may be described as
of fair size, deriving his distinguishing characters from his
head, horns, fleece, and general contour. The head is very
handsome with bent or Roman profile, and is decorated with
horns in both sexes. The head and short round ears are

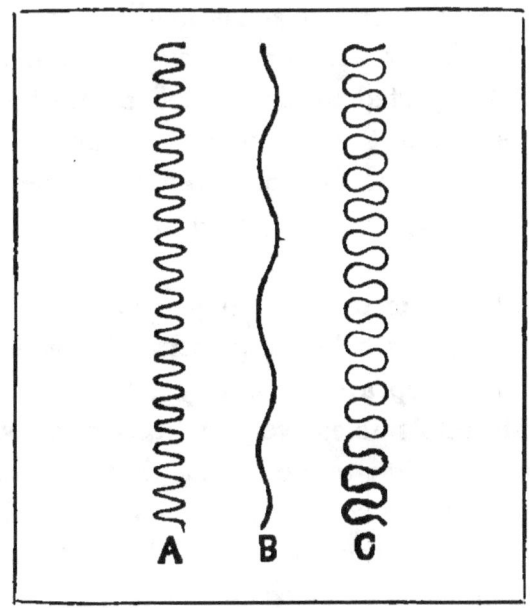

well covered with wool. The horns are open and wide
between, well turned, and marked with fine transverse
wrinkles. The nose is often pink, but a darker colour is
preferred. The neck is short, but full, and gains character
from the heavy folds (falken) of skin which adorn both males
and females. The shoulders are very broad over the tops,
but in some examples the withers are too high and pointed.
The body is long, the ribs deep and well sprung, and the
hindquarters are apt to droop. The legs are short and the
hocks incline to be narrow or cat-hammed. There are also

folds of skin gathered together over the tail, giving the puckered appearance known as the "rose." The general appearance of Merino sheep must strike an English sheep-breeder as quaint. The white face, pink nose, and light-coloured horns, the skin hanging in folds about the neck and rump, and the almost black outer surface of the fleece—all combine to produce a striking effect. The blackness of the fleece is, however, only at the surface, for on catching the animal and opening his wool with the thumb and two first fingers of each hand, it may be parted down to the skin as a funnel-like or lily-shaped orifice, showing wool of silk-like fineness, coloured with orange zones shading into pale yellow, produced by the fat or yolk of the wool. This very pretty appearance of the parted fleece effected neatly by an experienced hand is called the "blumen" or flower in Austro-Hungary. It may also be compared in shape to an old-fashioned wine-glass, the rim being at the surface of the fleece.

The greatest care is taken in the management of Merino flocks, not only to obtain but to preserve the wool in perfect condition. The sheep are constantly tended by shepherds, and are brought into stables every night, or whenever it rains. It is no uncommon thing to see the shepherd running towards the sheep sheds followed at a gallop by all his flock on the approach of a heavy shower. This close attention causes Merino sheep to feed in mobs, shoulder to shoulder. The peculiar dark colour of the extremities of the fleece is owing to dust and dung adhering to the greasy wool, which forms the almost continuous coating already mentioned.

The following measurements of specimens of Merino wool now lying before me, obtained from sheep at the Vienna Exhibition in 1873, will give a correct idea as to its general character :—

1. Rambouillet ewe wool, bred by C. H. Kayser, Haubitz, Grimma, Saxony, 2⅜ in. long, very dense and firm in fibre, resembling pale yellow silk.

2. Rambouillet ram wool 2½ in. long and of similar appearance to the above.

3. Rambouillet ram wool, 2¾ in. long, eleven months old, bred by F. Schwartz, Lappenhagan, Hohenfelde, Pomerania, Prussia.

4. Rambouillet ram wool, three years old, 2½ in. long, bred by C. H. Kayser, of Haubitz, Grimma, Saxony.

5. Cotswold-merino wool, 3 in. long, less dense and coarser, exhibited by the Keltschau Sugar Company, Moravia.

6. Prussian Rambouillet wool, 3 in. long and very fine, two years old, bred by H. Kannenberg, Gerbin Kosternitz, Prussia.

7. Fine Hungarian Merino, 1 to 1¼ in. long, from a three-year-old-sheep, bred by Count Alois Karolyi, Stampfen, Hungary. This wool is very fine and full of fat.

8. Hungarian Electoral-Negretti, 1 to 1⅛ in. long, bred by the Countess Laura Henckle, of Karlsbad, Sarndorf.

9. Ditto, ditto, from ⅞ to 1 in. long.

10. Cotswold-Negretti, bred by Count Frees, of Czernahora, 3 to 3¼ in. long and very hollow.

Specimens of Australian Merino also before me show a less perfect closure of the fleece, less fat or grease, and a considerably greater length of fibre.

1. An interesting sample of wool taken from a six-tooth ewe which had missed shearing one year measures nine inches long without stretching, and when stretched twelve inches. The fibre is perfect throughout, and, therefore, disproves the assertion so often made, and stated as a fact by Professor Low, that sheep in a state of nature cast their fleeces every year.

2. A sample of hogget wool 6½ in. long, and of fine fibre, resembling spun silk.

3. A sample of four-tooth wether wool 4 in. long, of nice quality, but wanting in the characteristic solidity of the Saxony wool.

4. Various samples of Australian Merino wool, about 4 in. long, and of fine appearance, but not so dense in the pile as the Saxony Rambouillet, nor yet so full of fat.

The contrast between these fine clothing wools and our home wools is most striking. The fibre of a Lincoln, Cotswold, or even a Southdown is coarse and thick beside the Merino. There is in our long wools an absence of curl or crinkle, the fibre resembling hair rather than real fine wool. It is loose, lashy and long, whereas the Merino is compact (solid), springy, and short. Lincoln or Cotswold wool 10 in. long will not stretch, whereas Merino wool of 4 in. will readily stretch to 5 in., and in one case above cited Merino wool of two years' growth measured 9 in., and when stretched 12 in. These are the carding wools from which dress cloth is made, and it is different in its properties from the long wools, which, as well as our native short wools, are subjected to combing, and destined for the manufacture of worsted yarns.

CHAPTER III.

BRITISH LONG-WOOLLED SHEEP.

LEICESTERS.

THE Long-woolled races of sheep are essentially English in origin. They are represented by the Lincoln, Kent, Cotswold, Leicester and Devon Long-wool breeds, but especially by the three first-named races. The position of the Leicester breed is particularly interesting, not so much on account of the length of its wool, which indeed is short in comparison with that of the other long-woolled sheep, as because of its historical connection with the improvement of all the other long-woolled varieties of sheep. The premier place is by common consent awarded to the Leicester on account of its having been the first breed upon which the skill of the breeder was applied.

The Leicester sheep appears to have inhabited Leicestershire and the neighbouring counties for a long period before it was subjected to improvement. Markham, whose twelfth edition of "Cheap and Good Husbandry" bears the date of 1668, writing of the sheep of the Midlands, says, "The sheep of Worcestershire, which joineth on Warwickshire, and many parts of Warwickshire, all Leicestershire, Buckinghamshire, and part of Northamptonshire; and that part of Nottinghamshire which is exempt from the forest of Sherwood, beareth a large-boned sheep, of the best shape, and deepest staple; chiefly they be pasture sheep, yet is their wool coarser than that of Cotsal" (Cotswold).

Professor Low also says: "There is no reason, therefore,

2

to assume from any of the characters presented by the wool of the new Leicester breed that the parent stock was any other than the Long-woolled sheep of the Midland Counties."

It was thus the ordinary sheep of his district that Bakewell used in bringing out his New Leicesters or Dishley breed apparently without having recourse to crossing, but by a confident reliance upon selection only.   Bakewell's success was due to a firm faith in the power of animals to transmit their good qualities to their progeny, and a steadfast adherence to the type which he wished to produce.   With him beauty of form, utility of form, early maturity, and good fattening properties were the main points to be secured, and he was apparently careless as to wool.   He was a positive, secretive, and self-reliant man, and left very little record of his proceedings.

That Bakewell deserves to be considered a genius there can be no doubt, and few inventors have merited higher praise than he, or have added more to the greatness and prosperity of their country.   The man who led the way to the improvement of all our breeds of live stock, and showed how in a few years the animal form might be modified, and improved incalculably beyond anything which had previously existed, surely deserves a national monument, and yet how little recognition has he received !   The name of Bakewell is rarely mentioned among those worthies who have made England.   It is not seen inscribed on the cornices of public halls, and his bust is missing in the corridors of the national pantheon.   Novelists, dramatists and painters, soldiers, engineers, and scientists may have their names recorded on the walls of fame, but Bakewell, the apostle of the art of improving the domestic animals of the world, is passed by unnoticed, and doomed to lie forgotten in his grave without a mark by which those who participate, as all do, in his grand work may know to whom they owe so much.   It is not out of place to call attention to this omission, and we cannot but think that the suggestion should receive attention, seeing that agriculture has now been raised to a position of greater eminence by the formation of an Agricultural Department of the State. The master-mind from whom Culley and Charles Colling derived their inspiration should not be lost sight of or allowed to be forgotten, for if the torch and fire of genius are to be claimed or accredited to any it is to the poineers of a new and great movement like this. A monument to Bakewell, flanked with effigies of the horses, cattle, and sheep which he ennobled, and adorned with a fitting inscription, would be as dignified in its silent teaching as that of a Stephenson, a Watt, or a Cartwright; and no place would be more suitable for its site than the metropolis of the country which he so well and unconsciously served. He was no seeker of or candidate for future fame, but as a farmer he worked without ostentation, until his influence spread itself over every civilised country of the globe.

Robert Bakewell, of Dishley, Loughborough, commenced the improvement of his county breed of sheep in or about 1755. The merit of his work consisted in his practically realising the fact that the properties of parents may be transmitted to their offspring until fixity of type is the result; also in his innate power of discerning by an animal's external form and "quality" that it possessed the properties he desired to perpetuate. He was able to discriminate between size and quality. He discerned the correlation between the kind of symmetry which he desired to see and aptitude to fatten—that is utility of form—and in this he evinced genius.

The result of Bakewell's work was the formation of an improved sheep somewhat less than the original type, but more symmetrical, thicker, deeper, and possessed of greater fattening properties as well as earlier maturity. Unfortunately the wool was neglected. His success as a sheep breeder is best indicated by the appreciation in which his animals were held. From a few shillings a head (it is stated that his first rams offered for letting only made 17s. 6d. each) the price rose to 100 gs., and in 1786 he made 1,000 gs. by the letting of his stock. In 1789 he made 1,200 gs. by the letting of three rams and 2,000 gs. for seven; and in the same year he made 3,000 gs. in addition, by letting the remainder of the rams to the Dishley Society, then instituted. These facts must appear extraordinary to anyone who reflects upon the greater value of money one hundred years ago than now, and the much less general appreciation at that time of the advantages of well-bred stock. Then there were no foreign buyers to stimulate biddings, nor princes and millionaires competing for favourite strains.

The Improved Leicester.

The Leicester sheep as bequeathed to us by Bakewell may be described as a white-faced, hornless race, covered with a fleece measuring about seven or eight inches in length, of somewhat lashy wool, but terminated with a short twisted curl. Describing its points briefly we find the following :— Lips and nostrils black, nose slightly narrow and Roman, but the general form of the face wedge-shaped, and covered with short white hairs; forehead covered with wool, although this is not always the case; no vestige of horns; ears thin, long and mobile; a black speck on face and ears not uncommon; a good eye; neck short, and level with back, thick and tapering from skull to shoulders and bosom; breast deep, wide and prominent; shoulders somewhat upright and wide over the tops; great thickness from blade to blade or " through the heart"; well filled up behind the shoulders, giving a great girth; well sprung ribs, wide loins, level hips, straight and long quarters, tail well set on, good legs of mutton, round barrel, great depth of carcase, fine bone, a fine curly fleece free from black hairs, well-covered back and loins, firm flesh, springy pelt, pink skin. The general form of the carcase square or rectangular; legs well set on, straight hocks, good pasterns, neat feet.

The Leicester is best fattened when from twelve to fifteen

months old, and the carcase then weighs about 80 to 100 lbs. There is an idea prevalent with some that the day of the Leicester is gone by. Pure-bred Leicesters are liable to lay on fat very thickly, and the demand for fat mutton has ceased, so that in this respect these sheep are somewhat at a disadvantage when compared with Downs. The great value of the breed lies in its excellent effects when used for crossing purposes, and it is premature to state that the Leicester is "played out." Good stock, although general, is by no means universal, and, besides, the Leicester cross produces wonderful carcases. What, for example, can excel the Southdown or Hampshire Down and Leicester cross? Again, what can be better for certain situations than the cross between Leicesters and Cheviots, or Leicesters and Blackfaces. Going further afield, the Leicester-Merino is an improvement upon the Merino, as it produces a butcher's sheep, as well as a mere wool-carrier.

There is scarcely a breed which has not felt the influence of the Leicester. No race has been so largely employed as a means of improvement, and Southdowns, Cotswolds, Lincolns, Shropshires, Hampshire Downs, and probably every sort of sheep, directly or indirectly has benefited by it.

Improved Lincoln Sheep.

In general form the Lincoln resembles the Leicester. An ordinary observer at a show might find it necessary to consult his catalogue to see where the Leicesters ended and the Lincolns commenced, but the differences are considerable, notwithstanding. Thus, in size, the Lincoln is greatly superior, and he may be credited with being the heaviest sheep of the British Isles, having been known to attain a weight of 90 lbs. per quarter. The flesh is firmer than that of the Leicester when tested by laying the hand upon the back. The wool is extraordinarily long, samples in my possession having been proved by measurement to be 21 inches in length, while the weight of a ram's fleece may touch 30 lb. The wool is very bright and lustrous when shorn, giving the name of lustre wool, a character which disappears when the sheep are bred away from their native county. The massing of the wool in flakes or strands upon the animal is characteristic of the breed, but the fibre is hair-like and " lashy " if separated and compared with the fibre of Merino wool. The history of the improved Lincoln sheep is very interesting. In order to understand it we must go back a hundred years to the controversy which then enlivened agricultural literature as to the respective merits of the old and the new Lincoln.

As already hinted the Leicester was the agent employed in the new departure and keen were the discussions and many the displays of temper before the momentous question was settled in favour of the new type.

As no one connected with agriculture has been more spoken of lately than Mr. Chaplin, it is interesting to find a Mr. Chaplin of those days in hot dispute with Bakewell in 1788, because the great Leicester breeder had dared to step in and have a look, unbidden, at Mr. Chaplin's rams. Mr. Chaplin wrote as follows : " After my refusal on the 21st inst. to let you see my sheep before they were collected and sorted at home, I did not expect to hear of your meanly sneaking into my pastures at Wrangle (most appropriate name) on the 24th, with two other people, driving my sheep into the fold and examining them." To which the great Bakewell rejoined : " We asked a young man if you had any rams there ; he informed us you had. ' Where are they ? ' ' In the close next the house.' ' May we see them ? ' ' Yes.' ' Who would show them ? ' ' I will.' From which we supposed he had frequently shown them to others. We then alighted and went into the close ; he opened the pen gate and we assisted him in driving them in, about fourteen in number." We have introduced this short extract from a rather angry correspondence as a little piece of human nature not without interest. It is life-like, and so similar to what happens even now when rival breeders take a sly peep at each other's proceedings as to be also instructive. Not quite dignified, perhaps, but truly showing that "one touch of nature makes the world akin."

The new or improved Lincoln is the product of Leicester crosses upon the old Lincoln. He is truly a magnificent creature of true long-woolled character, requiring rich pastures and plenty of space. The farmers of Lincolnshire are justly proud of their breed, and we see that Mr. Rew speaks of them as "this famous old breed," showing that the controversy of the closing years of the last century is forgotten. As a mutton sheep he is inferior to the Down breeds so far as quality is concerned, but for exportation for crossing purposes no class of sheep is in greater demand, and very high prices are paid for good specimens.

LINCOLN RAMS.

CHAPTER IV.

THE BORDER LEICESTER.

THE Border Leicester sheep is full of interest to north-country farmers. Not only has it achieved an independent position as a breed, but it has out-rivalled in its distribution, as well as in its value, the original Dishley stock from which it is descended. It is not too much to say that the Border Leicester is the mainstay of Border counties' farming. It is constantly crossed both with Cheviots and Blackfaces for the purpose of breeding wether lambs, and it is also well maintained as pure-bred stock over many large districts, while its improvement has been, and is, carried on with great enthusiasm and judgment.

At the Windsor Show the Border Leicester breeders appeared in equal force with the Leicester men—seven to seven—and those who stopped to converse with the shepherds and masters of the Scotch contingent must have been struck with their keen appreciation of points and their love of the race which they tend. Up to 1869, the year of the Manchester meeting, all Leicesters, whether belonging to the Midlands or to beyond the Tweed, competed in the same class—an arrangement which was found to be unfair, and since then they have been ranged under different banners. The great importance of the breed cannot be shown better than by the fact that at the annual Kelso sales upwards of 2,000 rams are disposed of, a muster which the Midland breeders would find it difficult to surpass.

History is always interesting, especially when it reveals the habits and the thoughts of earlier generations. Once more we must glance back into the last century, to the times of Bakewell. At that period the way for improved cattle and sheep was being prepared by the introduction of drill husbandry. Bakewell drilled his turnips, and Dawson, who was in service at Dishley, learnt the new system and carried it down to Frogden in 1763. The same year George Culley visited Dishley, his brother Matthew having found his way there the previous year, and an intimate friendship sprang up between Bakewell and George. The connection be-tween the county of Durham and Leicestershire at that time had most important bearings upon the whole future of live stock. The brothers Colling were about to commence the improvement of the Shorthorn race, and it is said that Charles was a visitor at Dishley, where the Longhorns were in full force.

The Culleys were settled at Denton, which was within a short distance of Barmpton and Ketton, and the Culleys and Collings must often have met at Darlington market, or in the King's Head. At that time the Teeswater breed was in high favour as a long-woolled sheep, and the Culleys brought Leicester rams from Dishley, and continued crossing with the Teeswaters until they established a flock of Leicesters. In 1767 the Culleys took Fenton, near Wooler, and after-wards Wark and other farms until they paid rentals up to £6,000 a year. There, they were on intimate terms with the Greys of Millfield, and I have often heard the late John Grey of Dilston, who was born about 1784, speak with great ad-miration of them. The Culleys' stock are looked upon as the original strain from which the Border Leicesters were derived. Their rams were hired by breeders both on the English and Scotch sides, among whom were Messrs. Robertson of Ladykirk; Thomson of Bogend; Thomson of Chillingham Barns; and many other noted men. It is in-teresting to find that the excellent sheep still held by Mr. James Thomson (the laird of Mongoswalls), and grandson of

the tenant of Bogend, are the direct descendants of the
Bogend flock. I can speak from personal knowledge of the
excellence of this old-established flock, which, being continued in the same family for upwards of a hundred years,
forms the best link between our time and the period of the
Culleys, in the history of the Border Leicester breed. The
Culleys retired in 1806, and the dispersion of their flock
became, through the flock of Compton of Learmouth, the
progenitors of Lord Polwarth's Mertoun flock, now awarded
the place of highest honour among the entire breed. This
famous stock was, it is said, begun at an even earlier date
by Mr. Scott, the then proprietor of Harden and Mertoun,
by purchases from Culley's noted flock, so that Mr. Scott's
sheep were spoken of as " not inferior to the sheep of Mr.
Bakewell." Mr. David Archibald, of Awa Moa, Octogon,
New Zealand, who has traced the history of Border Leicester
sheep with commendable pains, says that according to this
account the Mertoun flock was contemporary with the Dishley and Wark flocks. Lord Polwarth, however, claims the
more modest antiquity of 1802 when he states that the
Mertoun flock was started by his grandfather, Mr. Hugh
Scott of Harden, by purchases of ewes from Waddell of
Mousin, Burn of Millfield, and Mr. Robson, and to these he
afterwards added a number of ewes from Mr. Jobson's flock
at Chillingham, Newtoun. The early rams were hired from
the Culleys, to whom as much as 100 gs. was paid for their
use for a season. In 1888 Lord Polwarth sold twenty-eight
shearling rams at the Kelso sale for an average of £36 9s. 3d.,
and one for 165 gs., showing the extraordinary estimation in
which pure descent as well as high quality is held, and bearing
out the opinion, which has often been expressed, that there
is no limit to the value of high-bred stock if the correct lines
are adhered to. The average of 812 rams from twenty-one
noted flocks at the same sale was £11 7s. 10d. each. It may
not be out of place to mention that in the same season 138
rams of the English Leicester breed made an average of

£6 17s. each, the highest price given being £16 10s. In 1890 the average for Lord Polwarth's rams was £53 19s. 4d.

The differences between the Leicester and the Border Leicester are to be chiefly seen in the head, which in the Border Leicester is white, and boldly carried, the nose slightly aquiline, the muzzle full, the nostrils wide, and the ears erect. The head is clean and free from wool, as is pretty well shown by the fact that they suffer from flies settling on their polls in summer. The English Leicester, unless trimmed and shaved for show, usually carries a tuft of wool on his head, which protects it from flies, and he is also woolled in the shanks. The English Leicester has a bluish-white face; whereas the Border Leicester's face is clear white. In carcase the Border Leicester is the larger and longer, and the belly is not quite so full in outline, being carried rather more lightly. As to carcase points, these are so similar in all sheep that they may in the present case be omitted, especially as they have been given in detail in describing Leicester sheep.

The idea which has been expressed that Border Leicesters were produced by crossing Cheviots and Leicesters is not accepted by their breeders, and Mr. Archibald, who has been already quoted, scouts it as untrue. The well-known history of such flocks as the Mertoun, Mongoswalls, Mellendean, Bonnington, Mersington, and The Rock, are sufficient guarantee that Dishley was the original home of both English and Border Leicesters, and that the brothers Culley brought out the Border Leicesters by repeated crossings of Dishley rams upon Teeswater ewes. In Northumberland, which is the native county of the breed, they found patrons in the Bosanquets, at The Rock, in the Greys, of Millfield, and, later, in the Messrs. Dinning, the late Mr. John Atkinson, of Bywell Hall (Peepy), and Andrew Wood, on the banks of the Tyne.

From Stamford Bridge in Yorkshire, through the counties of Durham, Northumberland, and into Lothian, the Border

Leicester is the important race, and the fertile lands around and east of Carlisle also are stocked with the breed. A favourite cross for producing a hardy and prolific ewe flock is that of Border Leicester rams mated with Cheviot ewes, and these again put to "Leicester" rams give a strong and kind progeny, which may be brought out fat at a year old at from sixteen to twenty pounds per quarter. Such ewes are well adapted both for the richer lands which form the valleys, and the fells which enclose them. They are a frugal race, requiring but little water, and thriving upon the poorer pastures of the mountain limestone and the millstone grit. These sheep fatten easily without hay, and do not require so much indulgence in the form of cake and corn as some of our south country Down breeds. Oats and pea straw, with a pint of home-grown oats, and plenty of white turnips is all they require in winter, and in summer they have nothing but grass. The fault of the "bred ewe" or pure Leicester is that she makes herself too fat upon good land and hence the Cheviot cross or the Cheviot twice crossed is preferred for the outlying lands as hardier and better adapted for poorer soils. Many holders of hill farms who also occupy farms in more favoured districts keep two flocks, one of Leicesters, which they breed at home, and a Cheviot or half-bred flock on the uplands, which they work by judicious crossing. The lambs are reared on the hill sides, and change hands at St. Boswells, St. Ninian's, Lauder, Stagshawbank, Brough Hill, and other great north-country fairs. They are then put on turnips or finished. In other cases the breeder shifts them from the higher grounds, and "turnips" them himself, thus securing the profits of both breeder and grazier. The still higher and more inhospitable ranges are stocked with the hardy Cheviot or the still more frugal Black-faced or heath-breed, and thus all the districts of that varied and picturesque part of the kingdom find suitable sheep stock.

LIBRARY OF THE UNIVERSITY OF CALIFORNIA

CHAPTER V.

COTSWOLDS.

THIS breed, contrary to the habit of most long-woolled sheep, has made its home on bleak uplands. Visitors to Cheltenham, or residents in the fertile vales of Evesham, Gloucester, or Berkeley, may raise their eyes to the rugged outline of Cotswold, and see the hills covered with snow while the valley is basking in sunshine. The beautiful scenery of Warwickshire and Worcestershire is due in a great measure to the alternation of valleys cut deeply through the rocky foundations on which these hills rest to the smoother flats of the lias clay. Leckhampton, Quidhampton, Birdlip, Tetbury, Cirencester, Stroud and North Leach, Slaughter and Bourton-on-the-Water fairly indicate the area of this tableland, intersected with deep ravines, which compose the Cotswold Hills.

Cotswold sheep are among the most ancient of our recognised breeds. The hills take their designation from the sheep rather than the sheep from the hills. They derive their name from *cote*, a sheep fold, and *would*, a naked hill. On these woulds, says the translater of Camden, "they feed in great numbers flockes of sheep, long-necked and square of bulk and bone by reason, as is commonly thought, of the weally and hilly situation of their pasturage, whose wool, being more fine and soft, is held in passing great account amongst all nations." Stowe, in his Chronicles, states that in 1464 King Edward IV. " concluded an amnesty and league with King Henry of Castill and King John of Aragon, at the

concluding whereof he granted licence for certain Coteswold
sheep to be transported into the country of Spaine, which
have there since mightily increased and multiplied to the
Spanish profit." This and other evidence seems to point to a
fine-woolled breed more nearly resembling Merinos or Leominster sheep than the modern Cotswold, which is evidently
not the sheep alluded to in these old records. Professor Low,
writing at a period now belonging to a past generation, tells
us that the Cotswold sheep of to-day had inhabited the
district " beyond the memory of the living generation," which
would take us back far into the last century. It is probable
that they are an offshoot of the Midland long-woolled sheep
which Markham speaks of as extending into Warwickshire
and Worcestershire. And yet, in the middle of the seven-
teenth century, this authority says: " The sheep upon Cotsal
Hills are of better bone, shape, and burthen, but their staple is
coarser and deeper than the Lempster (Leominster) side."
And, again, speaking of the Leicestershire sheep, he says:
" Yet is their wool coarser than that of Cotsal," showing that
the Cotswold sheep at that time were intermediate in wool
between the Ryeland, which is the Lempster breed, and the
old Midland Leicester. The short pasture of the Cotswolds
would naturally tend to produce a finer wool than the rich
grass land of Leicester and Warwickshire, and hence it is
not difficult to account for the difference in the quality of the
wool. The writer was well acquainted with some of the old
standards of thirty years ago on the Cotswolds, and has heard
them speak of their fathers going regularly into Leicestershire
to buy rams, and there is no doubt that the old type of sheep
was greatly improved by the new Leicester. The breaking
up of the downs and the cultivation of the turnip could not
fail to affect the sheep, and would tend to increase the weight
of the carcase and the length and strength of the wool. The
attention of the Cotswold men was directed to wool of a
certain class, for they have long preferred a bold and open
curl rather than the close spiral of the Leicester. The face

of the Cotswold indicates a disposition to grey or light brown, and the same appearance is to be noticed upon the shanks, which may point back to crossing with the original fine-woolled race.

All Cotswold sheep men will remember Mr Smith's (Bibury) grey-faced Cotswolds, and that they were stated by their owner and breeder to be pure Cotswolds, without any admixture of Down blood. A few speckles of grey was not thought to be a drawback a few years ago, whatever the present fashion may be, and hence it is probable that the Cotswold sheep may have originally had a light grey face.

The Hewers, Lanes, and Garnes have long been associated with the improvement of the breed of their district, and at the dispersion of Mr. Hewer's famous flock at North Leach, the foundation of more than one prize-taking flock of the present day was laid. Whatever the origin of these sheep, it is certain that for the last sixty years at least they have been kept pure, so that the type is now fixed, and no Cotswold breeder need fear the imputation that his sheep are of a mixed origin, except in the sense that all our sheep have at some period been improved by crossing. The Cotswold sheep may be described as big and upstanding, and of better carriage than the Leicester or Lincoln. The late Mr. John Algernon Clarke expressed his opinion that a Lincoln sheep should have " no neck," by which he wished to convey the idea of a short thick neck, quickly blending with shoulders and bosom. The Cotswold was often described by the late Professor Coleman, when teaching at the Royal Agricultural College, as a sheep which could "look over a hurdle," that is, carried his head high and well poised on a somewhat erect neck. This is said to be accompanied with a tendency to be "ewe-necked" and low in the rumps, or "down at both ends." as I have heard Professor Coleman repeatedly say. He knew the Cotswold sheep well, and was himself a Cotswold man, so that his opinion carries weight. That these faults have been corrected in the best flocks there can be no doubt, but it exists in second rate-animals, and is most evident in rough weather, when the animals are viewed at a disadvantage.

The characteristics of these sheep are as follows:—The head is wedge-shaped, and the nose is straighter than that of the Leicester. The face is covered with white hairs, sometimes interspersed with light grey patches or specks. The lips are black, as is also the skin close to the eye. The ears are long and flexible, and inclined rather upward. The forehead is decorated with a flowing top-knot, which is never shorn or shortened, even when the sheep are being trimmed for show. This forelock must be considered as one of the most characteristic features of the race. The carcase points of all breeds of sheep are very similar, and much of what has been advanced with reference to Leicesters is equally applicable to Cotswolds. As the late secretary of the Farmers' Club pointed out, character lies in the head, and when the head is off there is little difference between the various long-woolled breeds. Thus good shoulders, and well let down legs of mutton, ample loins and well sprung ribs, depth and squareness of carcase, good girth and well carried out quarters, are formulæ which well might be repeated over every breed. We shall therefore assume that, so far as carcase is concerned, well-made sheep should possess all these points, and content ourselves with the consideration of those characters which really mark the race. The skin of the Cotswold is covered with a fleece of great length and weight, although inferior in both points to the Lincolns. In "crack" flocks the fleeces run three to the tod of 28 lbs. or 9⅓ lbs. each, and many fleeces have been shorn weighing close upon two to the tod; but this must be considered as exceptional.

The wool lies on the surface of the fleece in large round curls, and the fibre is somewhat coarse and hair-like when examined separately, and has no wavy wrinkles such as are seen in Shropshire or Hampshire wool. The sheep stands straight on his legs, and carries himself nobly.

Cotswold lambs are somewhat ragged in appearance during

their first summer, looking dry in the coat and wanting the bloom which is seen on Down lambs. They are said to be delicate when young, and require time to mature. But whatever their appearance in youth, they grow up grand and hardy, and it is asserted by experienced farmers of the district that no breed will so well withstand the climate or thrive so well on what Camden's translator calls the " weally " nature of the soil. They ought not to be run too thickly upon the land, nor enclosed too constantly in hurdles, and if such a policy is pursued they lose in size. The Cotswold is not adapted for breeding fat lambs. The mutton is of second quality, like that of most long-woolled sheep, and is pale and long in the grain. When long wool commanded a higher price than short wool, the position of the Cotswold sheep was stronger than at present, as a Cotswold fleece was easily worth £1. Now the same fleece is probably not worth more than 10s. For a time the demand for Cotswold rams seemed in danger of falling off, but during the last two or three seasons there has been a reaction in their favour.

THE KENTISH OR ROMNEY MARSH SHEEP.

The flat and bleak district of Romney Marsh would scarcely at first sight, seem suitable for extensive sheep farming. And yet there is a general resemblance between it and the extensive fens of East Anglia, which once supported the famous old Lincoln breed, and now carry its descendants, modified by Leicester crosses. A similar and parallel change transformed the large and hardy sheep of Romney Marsh into the compact race which to-day covers the large tract stretching from Hythe to the River Rother, and for ten miles from Dungeness to Appledore. Over the westerly district known as Romney Marsh proper, Walland, Denge, and South Brooks; from Hythe in Kent to Guildford Marsh on the north-west, stretching well into Sussex; we find the district known as Romney Marsh. It is a plain of alluvial land nearly on a level with the sea, and protected from the tide by sea walls, probably of very ancient date, as in the time of Henry III. their regulation and rights were spoken of as ancient and approved. The soil is usually a deep alluvial clay interspersed with portions of infertile sand or gravel, and the area is traversed by wide ditches full of water. Romney Marsh boasts a humid and scarcely salubrious climate, and carries a sparse population, most of whom are employed in tending the sheep, which are maintained in greater numbers than on any equal area in the kingdom.

I forbear to describe the ancient sheep of Romney Marsh in the uncomplimentary language usually employed by writers on live stock with regard to the unimproved races. One thing is certain, they had friends in the past as well as traducers, and it appears doubtful taste to slate a dead breed now unable to return the compliment or stand up for its big head, narrow chest, "flat sides," or "big belly"—all of which seem to have been characteristic features of the older races of live stock. The battle of the races appears to have been a stout one, although eventually decided in favour of the modified forms resulting from importations of new Leicester blood. Whether the old Romney Marsh really appeared as he is depicted by an artist following the descriptions still extant in Culley or David Low, is exceedingly doubtful: that is, with "broad feet, long, stout limbs, narrow chests, flat sides, and great bellies."

The apparently fanciful description of a Warwickshire ram, by Marshall, is certainly grotesque, and seems to be an unfair contrast between the past and present. "A frame large and remarkably loose, his bone heavy, his legs long and thick, terminating in great splaw feet, his chine, as well as his rump, sharp as a hatchet, his skin rattling on his ribs." Such a description would scarcely have commended itself to those who maintained that the old breed had its advantages, and who considered that the new type was not in all respects better than the old.

Writing upon the Romney Marsh sheep, Professor David Low records in his time that "it may be doubted if there now exists a single long-woolled sheep in the county of Kent in which the influence of the new Leicester blood does not appear." And yet, some years later (1855), his successor, the late Professor Wilson, wrote: "Attempts have been made at various times to introduce Leicester blood into the flocks, but they have not been altogether successful—the shape and points of the animal have been improved, and earlier maturity and aptitude for fattening obtained, while at the same time the size of the sheep has been somewhat diminished, and the fleece, though improved in staple, has been reduced in weight. It has also been found that if the Leicester blood predominates, or even exceeds a certain point, the natural hardihood of constitution is changed, and the sheep become too tender for their exposed pastures." This is careful writing, and no doubt a correct criticism upon the introduction of Leicester blood. Even the most recent of the two quotations given now belongs to a past generation; we see from it how the memory of the previous crossings spoken of by Low appeared to be forgotten. The Romney Marsh sheep is now bred with care on both sides, and it would scarcely be fair to speak of the breed of to-day otherwise than as a distinct variety. What Professor Wilson spoke of in his day as the "pure" breed had white head and legs, long and broad face, with a tuft of wool on the forehead; no horns; neck long and thin; breast narrow, with moderate forequarters; the body long, with flattish sides and sharp chine; loins wide and strong; the belly large; thighs broad and thick; and legs and feet large, with coarse bone and muscle. They are also described as very hardy, and as bearing closer stocking than other breeds.

This description scarcely fits the new breed, which, during the last thirty years, has undergone great improvement in all respects. The practice of naming sires and keeping a pedigree of good sheep is now adopted by many breeders with good results. Mr. Thomas Brown, of Marham, has introduced

the same system of recorded pedigrees in his flock of Cotswolds, and the same thing is to be noted in the Romney Marsh
flocks of Mr. Neame and of the late Mr. Thomas Finn. Care
in breeding is also evinced at Sharsted Court, where the rams
are kept apart and the ewes brought to them when in season
and mated according to points and peculiarities.

The Romney Marsh or Kentish Long-woolled sheep of our
day is of large size, and is in this respect scarcely excelled by
any British breed excepting the Lincolns. A pen of three
ewes exhibited by Mr. Henry Rigden, of Lyminge, at a
recent Show of the Smithfield Club, weighed 8 cwt. 2 qrs.
24 lb., which was only excelled by two pens of Lincoln ewes,
which were respectively 9 cwt. 1 qr. 22 lb. and 9 cwt. 2 qrs.
14 lb. The breed is white-faced, hornless, and inclined to
be bareheaded. The wool is of long staple and great
weight, and in general appearance the breed resembles the
heavy long-woolled Lincoln race. Fine specimens are to be
seen at our great shows, not only from the true Marsh district, but also from Sittingbourne and other localities of good
land, of higher position.

THE DEVON LONGWOOL.

This breed has been known for centuries in the neighbourhood of Bampton or Bathampton, a market town and parish
in the hundred of the same name in Devonshire. The old
town stands in a pleasant vale, and the houses are built of
stone, and are irregularly scattered over a space extending to
about half a mile, near the river Batherm. In *Bell's Gazetteer*,
published in 1836, we read " Many sheep are fed in the
neighbourhood; they are of a large size, and of an uncommonly fine quality, from the excellence of the pastures."
Twenty years later Professor Wilson wrote, " It is very difficult
to find the pure Bampton unmixed with other blood: a few
only remaining in Devonshire and West Somerset."

The original Bampton Nott was a large-framed, heavy-woolled sheep, white-faced and hornless. It appears to have been an out-lying branch of the long-woolled sheep of the country which spread from the Tees to the Severn, following the flat tracts and undulating ground of the new red sandstone north and west of the oolitic formation. It was no doubt from this breed, extending from the Tees mouth to Warwickshire and Worcestershire, that the new Leicester itself was derived, and, if this surmise is correct, the subsequent improvement of the allied long-woolled races by the Leicester would scarcely be a cross, but rather the introduction of a selected strain of a similar origin to their own. Whatever view we may be disposed to adopt, there is no doubt that the old Bampton breed has been modified by repeated Leicester and Lincoln crosses.

The Devon South Hams were found from the Vale of Honiton to the borders of Dartmoor. Originally they had brown faces and legs, but in other respects resembled the heavy Romney Marsh breed. These sheep have also been improved by the introduction of Leicester blood, which has reduced their size, and has caused the dark colour of the faces and legs to disappear.

The Devon Longwool are now an established race, and were well represented at the great meeting of the Royal Agricultural Society in 1889. These sheep are the present representatives of the old Bampton breed, which does not appear as a distinct breed in the schedule of the Royal. Specimens of this breed exhibited at Islington showed distinct evidence of their Leicester and Lincoln origin, being stronger in type both as regards carcase and fleece than the former type.

ROSCOMMON SHEEP.

These sheep have been naturalised in Ireland for a long period. They are described by Culley in the usual uncomplimentary language employed by writers towards unimproved races generally. A vast improvement has been effected by crossing with the Leicester breed, and the Roscommon sheep of to-day is an exceedingly well-made animal, with the characteristic white face and wool of the Leicester. Anyone who reads the description of the Irish Long-woolled sheep given in his time by Professor Low, and compares it with what he may himself see in Ireland, will come to the conclusion that the breeders of these sheep have not been idle during recent years. The soil and climate of Ireland are favourable for the development of the larger and longer-woolled sorts of sheep, and in the Roscommon breed we see a good instance of successful improvement.

THE WENSLEYDALE LONGWOOLS.

These appear to be the modern form of an old breed once well-known as Teeswaters. They are closely allied to the Leicesters, and it is claimed for them that they, as well as the old Lincolns, were employed by Bakewell in his work of selection and improvement.

The great similarity in type between the Wensleydale, Leicester, and west country Devon Longwools is very striking, and evidently points back to a common origin. These Longwoolled races of Leicester type appear to have occupied the country from Yorkshire on the north-east to Devon and Somerset on the south-west, and to have extended through Nottingham, Leicester, Warwick, Worcestershire, and Gloucestershire, and adjoining counties. They therefore occupied a broad band in the Midlands and in the north-eastern and south-western counties, following, indeed, the northern boundary of the chalk formation, south of which the Down breeds have their home, and north of which the hardier mountain races of Wales, Derbyshire, Lancashire, Cheviots, and the Highlands of Scotland are to be found.

Mr. J. Heugh, of Mudd Fields, Bedale, writes to us as follows :—"The Wensleydale Longwoolled sheep appear to have had their origin in the valley of the Tees, on the borders of Yorks and Durham; as, up to recent years, they were known by the name of Teeswaters, and, in some districts, are known by that name to this day. According to some authorities, they took a part in building up the new Leicester or the Dishley sheep. Speaking of Robert Bakewell, and of the old breeds he used in his experiments, and in the proportion he employed them, one writer says :—' The old Leicester breed, which might come most readily to his hand, were large, coarse animals, with an abundant fleece, and a fair disposition to fatten, and they probably contributed not a little to his results. But other Long-woolled breeds, particularly the old Lincoln and two other breeds, respectively Warwickshire and the valley of the Tees, are also reported to have been more or less in requisition.' They took the name of Wensleydale Longwools about the time the Yorkshire Agricultural Society commenced giving prizes for them, being more extensively cultivated in Wensleydale than in other parts. They are more symmetrical, with a greater aptitude to fatten than formerly, the improvement being due to a cross of the Leicester; either from the noted Blue Cap, bred by Mr. W. Sonley, of Lund Court, Kirkby Moorside, or some of his descendants. The present type has long been fixed, and no one who has achieved any success as a Wensleydale breeder has deviated from a line of pure breeding. The Wensleydale is a large, high standing sheep, with a characteristic blue in the skin of the face and ears, but which sometimes extends to the whole of the body, though the shade is deeper on the face and shanks. The dark colour is cultivated, because of the extensive use of the rams for crossing with the Black-faced Mountain ewes; it is found that dark blue rams throw dark faced lambs—a point much valued. The lambs bred in this way are called Crosses in Scotland, where they are extensively bred. In the east of Yorks, and in Lincolnshire, where many thousands of them are annually bought and fattened, they go by the name of Mashams. The wool of the Wensleydale is of a uniform open character. Long silky locks should cover nearly the whole surface of the body, including the forehead, between the eyes, round the ears, and on the belly and scrotum. The hind legs down to the hoofs, and even the fore legs at times, have downy wool on them. Hairy wool on thighs is objectionable. The head is of good size and well carried on a long and strong neck, giving much greater style than is usual in most breeds. The nostrils should be wide in the ram, the back of the head flat, and the ears large and thin, well set on and well carried. The breed is specially noted for the absence of patchiness or excess of fat. The quality of flesh produced, the hardy constitution, and active disposition, enable it to maintain the position of first favourite in many districts."

CHAPTER VI.

MIDDLE-WOOLLED SHEEP.

IN introducing a new class of sheep it is necessary to report progress. We have hitherto dealt with the long-woolled
races of sheep, as illustrated in such breeds as the Lincoln,
the Leicester and the Cotswold. As mentioned earlier in
this Handbook, we have in England no really short-woolled
sheep carrying fine wool. The true short-woolled sheep is
the Merino of Spain, Hungary, Germany and Australasia.
These sheep are not bred for mutton but for wool, whereas
under the cloudy skies of Britain, and the abundant pasturage
which our climate favours, we find sheep of larger frame,
thicker flesh, and longer fleeces. These are the true long-
woolled and middle-woolled sheep, in which undoubtedly we
excel in a marked degree.

By contrast, it is true, the Southdown may be spoken of as
short-woolled; but, if we take into consideration the still
shorter wool of the Merino, we shall see the propriety of
regarding our " short-woolled " sheep rather as middle-woolled
in character.

The distinction is, however, very marked, as anyone can
see in visiting the various classes of sheep at any of our great
shows. The middle-woolled sheep, as we prefer to speak of
them, were originally found in every county of England, and
were represented by the old Ryeland, old Cotswold, and
other allied breeds, many of which have ceased to exist.
There, however, appears to have always been a class of black
or brown-faced sheep indigenous to the chalk districts of

England. Thus, if we follow the large area occupied by this great formation, we shall (with the exception of the Yorkshire and Lincolnshire wolds) find it in Norfolk, and with it appears the Norfolk Down sheep, short-woolled and dark featured.   There does not seem to be a county in which extensive downs of chalk occur where there is not a corresponding breed of brown-faced sheep. Suffolk, for example, has long boasted a Down breed, which extended into Essex, so far at least as did the chalk. Cambridgeshire, Buckinghamshire, Bedfordshire, Berkshire, Oxfordshire, Hampshire, Wiltshire, Dorsetshire, Surrey, Sussex and Kent—each and all boasted these active and hardy breeds of Down sheep. From these, the native counties of this particular class, they extended into others, and now find their home in Shropshire and the adjoining counties under the name of Shropshire sheep.

The leading differences between these and the long-woolled sheep have already been pointed out, but before treating at length each of the short-woolled breeds, it seems necessary again to draw attention to the great differences which mark them as a class.

Reference to Youatt's treatise on sheep, which appeared in 1837, would show the reader the great advance which has been made in sheep-breeding since his day. Many of the differences between the breeds of various counties have disappeared, and many new types have appeared. The Oxford, Shropshire, and improved Hampshire Down were then unknown in their present forms, and the Southdown and Leicester were then regarded as paramount. What a change has been wrought during these fifty years ! An inspection of the pens of these two breeds at Islington, at one of the recent shows of fat stock, showed these, once predominating breeds, in a minority. Not only so, but time had told upon them, either actually or comparatively. The Leicesters and the Southdowns exhibited symptoms of being over-bred. Their small frames, fine bone, puny faces, show too clearly the effects of close breeding carried out for over a hundred years, and contrast with the robuster forms and stronger heads of the more modern pure and crossed breeds.

We shall during the next chapters endeavour to trace the history, and point out the qualities, of the middle-woolled or Down races of sheep, and in doing so a brief notice must be allowed of the finest-woolled sheep which England has produced—the old Ryeland breed. Whether this breed was originally derived from the Spanish Merino or not, we cannot say ; but certainly no race is more deserving of the name of a fine-woolled sheep than this. It differs from most of our Down races, although there is a certain resemblance between it and the Dorset Horned race. It was, and is, white-faced and polled, and its natural habitat appears to have been that light-land district around Ross in Herefordshire, named the Ryelands, from its suitability to the growth of rye.

LIBRARY
OF THE
UNIVERSITY
OF CALIFORNIA

THE OLD RYELAND SHEEP.

The Ryeland sheep is one of those races which, having been superseded by others for a long course of years, seems now to have found friends and promoters. Twenty years ago the Ryeland sheep was spoken of as something that had been, but was no longer. Like the Longhorn among cattle it had found a rival, and that rival was the Cotswold. Thus, in 1800, Herefordshire contained 500,000 short-woolled sheep (Ryelands), furnishing 4,200 packs, the weight of the fleece being 2 lbs. In 1828 the number of packs of short wool had diminished to 2,800, but no fewer than 5,550 packs of long wool were grown in the county (Youatt). " This fact," says the author just quoted, " speaks volumes as to the revolution that is going forward, and plainly points out the farmers' interest and duty."

The old Ryeland sheep was the nearest approach to a fine-woolled sheep which we possessed. Its wool was coarser, certainly, than the Saxony Merino, but it was much finer

than that of the Southdown. Taking as an indication of
fineness the number of serrations and the diameter of the
wool fibres, we find that these three breeds compare as
follows :—

	Number of serrations per inch in length.	Diameter of each fibre. Inch.
Saxony Merino ...	... 2,720	1-840th.
Ryeland (old breed)	... 2,420	1-750th.
Southdown... ...	... 2,080	1-660th.

The Ryeland of the early years of the century seldom
exceeded 14 or 16 lb. per quarter in the wether, or from 10
to 13 lb. in the ewe. They had white faces and were polled ;
the wool grew close to and sometimes covered the eyes. The
legs were small and clean, the bone light, the carcase round
and compact and peculiarly developed on the loins and
haunches. It was particularly frugal in its fare, and would
endure privation of food better than any other breed. Sir
Joseph Banks, who was well acquainted with their constitution and habits, used to say that the Ryeland sheep deserved
a niche in the temple of fame. The weight of the fleece
rarely exceeded 2 lbs. A noted ewe bred by Mr. Welles, and
figured in Youatt's fine treatise on sheep, weighed when fat
25 lb. per quarter. The similarity of the old Ryeland breed
to the Merino, and a certain foreign appearance, favour the
idea that it was imported from another clime.

The breed appears to have extended over Herefordshire,
Monmouthshire, Gloucestershire, Shropshire, Staffordshire,
and to have been scattered from the Tyne to the Thames.

The practice of cotting the Ryelands, although now forgotten, was very ancient. During the winter and especially
at lambing time, they were shut up at night either in an unoc-
cupied building or in a place erected for the purpose. They
were then fed with hay or barley straw, or pease haulm given
to them in racks, frequently suspended by ropes, and so contrived as to be easily raised in proportion as the dung accumulated below; for neither the owner nor the shepherd

thought of cleaning out the place while there was room for the sheep to go in and out.

The reasons generally assigned for this practice of cotting were that the wool, being preserved from the weather, was kept sounder and finer; that the sheep were preserved in better health; that they were preserved from liver rot; that fewer lambs were lost in yeaning; that great losses were always experienced when the sheep were folded in the open field; and that much valuable dung was accumulated by the practice. The system of cotting has now been entirely discontinued in Great Britain, but those who have visited the Merino flocks of Southern Europe will remember the care which is taken to house the flocks, not only in winter, but in wet weather—the object being the growth of fine wool. It is also well known that rye and barley straw produce a finer wool than can be grown on a fuller diet of roots and cake. England is now well supplied with the finer descriptions of wool from her colonies, while the production of mutton has become a much more important consideration with flockmasters. The strength of the position of sheep, as a farm stock, also rests upon the fact that they trample and render firm our light lands, and at the same time manure them in an economical manner without the aid of the dung cart. The system was, therefore, early doomed, with all its supposed advantages, some of which still carry a lesson. It seems, for example, to have been noticed that ewes and lambs were less subject to mortality during the critical period of lambing, and this fact may well be taken to heart by sheep breeders. Without returning to the old system of cotting, it may be useful to remember that ewes and young lambs require more shelter and succour than they are usually allowed.

The New Ryeland Sheep.

The Ryeland breed of sheep is now again in the ascendant, but in a modified form. There is little doubt that the change

has been effected by crossing with the Leicester, but it is
only fair to existing breeders to say that for the last fifty years
no such cross has been employed. The carcase weight has
been considerably increased, and in place of the 14 lb. or 16
lb. per quarter already mentioned as representing the weight
of wethers, we now find lambs of ten months old arriving at
18 lbs. per quarter, and shearlings of sixteen months old
scaling 20 lb. and 22 lb. per quarter. The wool also, if it
has lost in fineness, has gained in weight, and a well known
flock (the Brook, Colwall) have averaged 8 lb., and made
10½d. per lb. unwashed. These are startling changes when
contrasted with earlier descriptions of the capabilities of the
breed.

And yet the Ryeland sheep preserves something of its
original qualities. It fattens with great rapidity, and is thus
well adapted for breeding fat lambs. It is still a white-faced,
hornless, short and close-woolled race, distinct from any other
breed. It may be compared to the Shropshire without the
black face and legs, and it is probable that Ryeland blood
exists in the veins of that now famous breed. There is also
still a certain resemblance to the Merino in the character of
the wool. Mr. Frank Shepherd, who is well known in connection with the breed, and whose father was one of the few
who, through evil and good report, stuck to the Ryeland
when he was at a discount, writes as follows :—

" The Ryeland, as you are doubtless aware, is one of the
oldest of British breeds of sheep, and some fifty years ago
was the leading breed in this district (Malvern). A desire for
new breeds springing up, it was allowed almost to become
extinct, but by a few good old judges refusing to part with
their stock for other blood the breed has been saved its existence. I have said ' a few good judges '; I believe I might
confine it to one man, for had it not been for my late father
I feel positively certain that the Ryeland in reality to-day
would be unknown.

" To my mind it is one of the best all-round breeds we have, producing mutton and wool of the finest quality, with great constitution, invaluable for crossing purposes. It is a well-known saying in Hereford market that 'no sheep will get fat lambs like a Ryeland ram.' I often hear old breeders lamenting over giving up their Ryeland flocks. Within the last ten years, however, the disposition to breed them up again has considerably increased the flocks of this county (Herefordshire), and, in Brecknockshire especially, Ryelands are fast on the increase. One thing greatly in their favour is they seldom suffer from foot-rot. The Ryeland of to-day is a much heavier sheep than was the case some thirty or forty years ago, and arrives earlier at maturity—compact in form, straight back, sides and underline, on short, well-set legs, the last white, as also the face, thick scrag, and head well covered with wool. I notice in the Journal of the Royal the judge's report of Ryeland wool at Windsor is anything but reliable. If readers of the Journal conclude by such report that Ryeland wool in general is as stated there, they are very much mistaken. No better wool is grown on any sheep. I admit the failure, however, at Windsor. There were only three exhibitors. My own wool ought never to have gone. In the first place, all my lambs were shorn the previous year, so that I had no teg wool, and in 1889, as an experiment, my sheep were unwashed; both these things would, no doubt, tell considerably against it in competition. I have on several occasions beaten the Shropshires in open competition, and as a hardy farmers' rent-paying sheep I have no hesitation in giving the preference to the Ryeland."

THE SOUTHDOWN.

Having had the privilege of knowing the late Mr. Ellman, and hearing from his own lips the story of the rise and progress of the Southdown sheep, and of the points and special merits of the breed, I feel as though brought into connection with its earliest annals. Upwards of a quarter

of a century ago the younger Ellman was a venerable old
gentleman of benign appearance. His memory extended
back to the beginning of the century, and he was able to
narrate all the particulars of his father's work in bringing
out the Southdown race, in the time when George the Third
was young. I heard him speak of the old-fashioned and
unimproved Sussex Down, as of small size and bad shape,
long in neck, low at both ends, light in the shoulders,
narrow at the fore-end, and shaped "like a soda-water
bottle," small in front, and heavier in the middle, large in
the bone, but boasting a big leg of mutton. The fleece
was not so close and firm as now, and the almost proverbial
expression of "four-year-old Southdown mutton" was more
applicable than in our days.

Mr. Ellman laid stress upon the great improvement effected
in the neck and forequarters by his father, and held that the
neck should be bold. It should rise high in the crest, and be
muscular and thick. The shoulders, he said, should be wide,
and this width should be maintained by well-sprung ribs,
great girth, grand loins, straight and ample quarters, and
good dock. The leg must be well filled inside and out, and
"as round as a cricket-ball." The fleece must be "board-
like" in its firmness, and show cracks down to the skin, as
the animal turns, presenting a firm and springy appearance.
Arthur Young saw the Glynde flock in 1776, and says, "Mr.
Ellman's flock of sheep, I must observe in this place, is unquestionably the first in the country, the wool the finest,
and the carcase the best proportioned; both these valuable
properties are united in the flock at Glynde. He has raised
the merit of the breed by his unremitting attention, and it
now stands unrivalled." The attention of the Farmer King
was early directed to the Glynde flock, and it is not im-
probable that the Royal patronage conferred upon the elder
Ellman was the original cause of the Southdown being taken
up by the reigning family, as well as by many of our nobility.
The Southdown has been spoken of as a "gentleman's"

sheep on account of its beauty, its adaptability for park life, and its extraordinary quality of flesh, especially if the sheep is allowed to attain the age of four years. The mutton is close grained, dark in colour, tender, juicy, and yields a rich gravy. The joints are not too big, and the proportion of lean meat is large. The Southdown has also been spoken of as a "butcher's sheep," because it lays up a large proportion of loose or inside fat and suet. It proves, or dies well, and may be relied upon to scale heavier than appearance warranted before slaughtering. The Southdown sheep boasts a purer lineage than most of our improved races. It has been alleged that a Leicester cross was made, but, according to Youatt, this cross was a failure, as well as that attempted with the Merino, so that the Southdown stands out prominently as a breed improved by pure selection, and not from crossing.

It is one of the indigenous races peculiar to the chalk hills of the southern counties, and appears to have existed for long upon that breezy upland which finds its most southern termination at Beachy Head, and runs inland, as a protecting rampart, north of Brighton, Worthing, and Arundel. The true Southdown sheep appears to have been confined to the Southdowns—a Southdown of the Southdowns—and to have given way to a larger and looser-formed animal, as the chain of chalks, on which it browsed, passed into the neighbouring county of Hampshire.

Its character has no doubt been influenced by its surroundings. Originally it was horned, but these appendages have long disappeared except as slugs, or accidental buds. Its light forequarter appears to have been due to the greater muscular development of the hind-quarters required in mounting steep slopes, and its fine short wool to the scanty and short herbage produced on the chalk. The fine grain of the mutton was probably induced by slow growth, and the peculiar nature of the pasturage. Like some other breeds, the Southdown perceptibly alters when transplanted into

4

other districts.   The frame inclines to increase in size, the
face often becomes darker, and the fleece heavier, and looser.
Breeders of such transplanted flocks find it necessary to
return to Sussex to buy rams, in order to prevent deterioration, or alteration, of type, and this becomes a source of profit
to those who breed them upon their native soil.

Taking the last thirty years as a period for comparison,
the disposition has evidently been to cultivate a lighter colour
of face than formerly, many of the exhibits at our shows now
being exceedingly light, and scarcely of the colour of the
fallow deer—which was Mr. Ellman's ideal as to the colour
of the head of a Southdown.   The head is small, and the
features refined to delicacy, and in some case inclined to be
dish faced.   The carcase is faultless, being beautifully drawn,
oval on the top, from set-on of the neck to the tail, thoroughly
filled up behind the shoulders, and square and massive from
chest to twist.

The older Ellman described the head as small and hornless; the face speckled or grey; the lips thin, and the space
between the eyes and the nose narrow, the forehead well
covered with wool between and around the ears.   The head,
according to the younger Ellman, is deer-like, but not too full
at the orbit or eye-cap, as such bony protuberances are found
dangerous in lambing ewes.   The ears of the Southdown are,
comparatively, short and rounded, and carried more erect
than in the Hampshire Down.   This gives an appearance of
greater narrowness between the ears.   The chap, or under
jaw, is fine and fleshless, assisting to produce the deer-like
appearancce already mentioned; and the eye is full and
placid, giving the same impression.   The appearance of the
Southdowns is characteristic and unmistakable. Their fineness,
smallness, and high-bred character contrast with those more
robust races, the Hampshire, Shropshire, and Oxford, and
it is this contrast perhaps which gives the idea of breeding
and fineness which strikes an observer in passing a number of
classes of sheep in review.

One of the advantages of the breed, which it shares with other Down races, is its power of resisting the evil effects of over-stocking. As far back as 1788, Arthur Young writes:— "Mr. Ellman, on 500 acres, has 700 ewes, lambs, and wethers in winter, and 1,450 of all sorts in summer, besides 140 head of cattle." These numbers read even now as extraordinary, before the days of cake feeding, or importation of cheap corn from America, and are scarcely to be rivalled. Cotswold or Leicester breeders would consider such a stock as inconsistent with keeping up either the stature or the health of their flocks, and yet such is the character of the Southdown that they can resist the in fluences of over-crowding. The number of sheep kept by Wiltshire and Hampshire sheep farmers in these days are not much short of those maintained by Mr. Ellman, and when we take into consideration the greater size of the Hampshire sheep, the weight per acre must be considerably greater.

The Southdown race in its own district was long well maintained by the late Mr. Rigden, of Hove, near Brighton, and by Mr. Penfold, of Selsey. This flock, which had existed for a century, was only recently dispersed, with what excellent result will be fresh in the memory of readers and breeders. The late Jonas Webb, whose honoured name is still held in the highest respect, was a fitting successor to the Ellmans, although farming on the downs of Cambridgeshire, and it was through him, in a great measure, that the true type was handed down to our generation. Mr. S. M. Jonas, of Chrishall Grange, Royston, and Mr. Henry Webb, of Streetly Hall, whose sale was a landmark in the history of the breed, possessed flocks descended from the older Babraham Southdowns, and both of these last-named gentlemen are relatives of the more famous Jonas Webb, whose name is worthy of special notice as a master in the art of breeding. That was a pretty anecdote, and literally true, of Jonas Webb, when exhibiting at a great Parisian International Show held under the auspices of the late Emperor. Napoleon III. paused to admire Mr. Webb's Southdowns, and inquired who the sheep belonged to. "Yours, your Majesty," promptly replied Mr. Jonas Webb, "if your Majesty will accept of them." The gift was accepted ; and a splendid present of solid silver plate came shortly afterwards from the Tuileries to Babraham.

The Southdown has been honoured by the patronage of a number of Royal personages and nobility, headed by H.R.H. the Prince of Wales, the Duke of Richmond and Gordon, the Duke of Hamilton, Lord Walsingham, the Marquis of Bristol, Viscount Hampden, Sir William Throckmorton, Colonel Sir Nigel Kingscote, the late Mr. Henry Brassey, Mr. Colman, M.P., and many other well-known members of Parliament and county men.

The Southdown has ever been a favourite with the rich, but it also deserves the patronage of rent-paying farmers. Still, the fact remains that weight of carcase and weight of wool, rather than the highest quality, have been for long more profitable and more rent-paying investments for the ordinary farmer, and hence the Southdown is less often seen in farmers' hands than in those of landlords, but there are large districts where the Southdown is the breed of the country.

For crossing purposes the Southdown has been particularly useful, and his blood exists in every improved Down flock. It is to the admixture of Southdown blood, made long years ago, that the improved Hampshire Down owes his position. Through the Hampshire Down the Oxfords get the same qualities, and the Shropshires were no doubt crossed with the Southdowns, as we shall see later. As the Leicester lies at the root of the improved long-woolled races, so the Southdown is the original, and was the first improved, and first short-woolled sheep used in the renovation of the dark-faced Down breeds. They have found their way into every county, and they have been, and are, exported yearly to keep up and improve the flocks of foreign countries.

CHAPTER VII.

THE IMPROVED HAMPSHIRE DOWNS.

THE improved Hampshire Down sheep is one of the more recent additions to the agricultural wealth of this country. It was only at the first meeting of the Royal at Salisbury in 1857 that Hampshire sheep were awarded classes to themselves, and some of the earliest improvers of the race are still living—notably, Mr. James Rawlence, Mr. William King, of Old Hayward, now ninety years of age, and his brother, Mr. Stephen King. No breed has made more rapid progress, either in absolute improvement or in the wide appreciation in which it is now held, than this. I trust my readers will excuse me if I dwell on this particular race of sheep at somewhat great length. I do this from no wish to give it special prominence, but because I have examined into its early history, and find that there is a good deal of material which has not yet seen the light. David Low lived before the birth of the Hampshire Down, and William Youatt wrote before the breed had been constituted. The late Professor Wilson wrote on sheep before the Hampshire Downs became prominent, and Mr. Rowlandson only gives them a brief notice. I therefore feel it not only due to the early improvers of the breed, but to the breed itself, to place on record a detailed account of its history, and trust that some of the information I am able to give will prove interesting to breeders.

THE OLD HAMPSHIRE AND WILTSHIRE RAMS

have long been merged in the present improved breed. The Hampshires originally were horned, tall, light, and narrow in the carcase, and usually with white faces and shanks. The Wiltshire sheep were originally known as "crooks," so called from the shape of the horn, which turned back behind the ear, and bent over the cheeks. They were the largest breed of fine woolled sheep in this country (*British Farmers' Magazine*, 1830). A Wiltshire lamb which weighed 24 lb. per quarter, and contained 14 lb. of loose fat, is described in the *Commercial and Agricultural Magazine* for April, 1800. "These sheep," says Youatt, "not only prevailed upon the Wiltshire Downs, and were much, and deservedly, valued there, but were found in considerable numbers in North Devon, Somersetshire, Buckinghamshire, and Berkshire. They were a peculiar breed, differing in the shape of horn, and in other points, from the sheep of any other part of the kingdom, and were probably indigenous to the Wiltshire Downs. If they were rather slow in feeding, they were excellent folding sheep, and enabled more corn to be grown in Wiltshire, in proportion to its size, than in any other county in England. . . These Wiltshires have now (1837) passed quite away." They were crossed " again and again " with the Southdowns, until every trace of the old breed disappeared, and a useful variety of the Southdowns remained—only distinguished from the true Sussex sheep by somewhat larger size, *lighter* colour, and a lighter and finer fleece. The last flock of the old Wiltshire horned breed disappeared about 1819 (Rawlence).

According to Mr. E. P. Squarey, " the Hampshire breed originated in a cross between the old Wiltshire horned sheep, as well as the Berkshire Knot, with the Southdown. From 1815 to 1835 the Downs of North Hants and those of South Wilts were very different. The Wiltshire Down was larger, perhaps less handsome, and not so uniform with respect to colour as those of Hampshire, and a ewe with a speckled face and ears was not always drafted."

About 1829, Mr. John Twynam, as he himself afterwards related to a farmers' club, used Cotswold rams. His idea was to blend together the best breeds then in existence, and by using an improved Cotswold sheep upon Hampshire ewes, he considered that he obtained an animal which united the qualities of the old Wiltshire, the Southdown, the Cotswold, and, indirectly, the Leicester. Mr. Twynam thus expressed himself:—" You must have observed an immense improvement in the character of the Hampshire sheep generally (185—), within the last fifteen years. I have had my attention called to this fact frequently since I have ceased to be a breeder. How has this altered character been obtained? Can we recognise none of the Cotswold fleece, or his more symmetrical proportions? And, when I tell you that in the years 1835-36 and subsequent years, I sold very many half-bred lambs, and not only into Hampshire Down flocks generally, but into those of six or eight of our first ram breeders, whose names are to be seen at this day upon my books; when as you must be aware these breeders are in the constant annual habit of selling one to another in this and adjoining counties; I trust I may, without presumption, lay some little claim to having supplied a portion of the material from which our present flockmasters have worked up a better and more valuable fabric."

In 1835 the sheep of both counties, and also of parts of Berkshire were, to speak generally, modified Southdowns, retaining some of the features of the older breeds—especially those of size and quality of wool; but had not arrived at the distinction of being a recognised breed. They were exhibited at the first show of the Royal Agricultural Society of Oxford in 1840 as West Country Downs, a name they long retained, and were at that time something like the present sheep, but smaller, looser, narrower at the fore-end, and lighter in colour.

It has been noticed that the crossing with Southdowns had been going on for many years before the formation of a breed was accomplished. Just as Shorthorns existed before the days of the brothers Colling, and Leicester sheep before the days of Bakewell, so Hampshire sheep had taken their general form before the days of Mr. Humphrey, of Oak Ash. Not only so, but Mr. Humphrey had contemporaries and customers from the first. Mr. James Rawlence (of Bulbridge), Mr. Stephen King, Mr. William King, Mr. Moore (of Littlecott), Mr. Edward Waters (of Stratford-sub-Castle), Mr. Frank Budd (near Whitchurch), Mr. Saunders (of Watercombe), Mr. Canning (of Chisledon), Mr. Ferris (of Manningford, Upavon), Mr. Bennett (of Chilmark), were all engaged in breeding these sheep. Mr. Humphrey is, however, by common consent looked upon as the man who lifted the sheep into its present position. Thus Mr. Squarey writes:—"To Mr. Humphrey, of Oak Ash, is due in a great manner the present character and position of the Hampshire Down sheep. This agriculturist effected its improvement by careful crossing with the largest and best fleshed of the Babraham Southdown flock. This means, applied with wonderful ability, and at a great cost, at length resulted in the present perfect animal." This justifies me in giving Mr. Humphrey a first place, while Mr. James Rawlence, the oldest living breeder (with the exception of Mr. William King, now retired), must be looked upon as his most worthy and best-known successor.

Mr. Humphrey, in a communication to Mr. W. C. Spooner, in 1859, gives a short account of the manner in which his flock of ewes was got together:—"About twenty-five years since, in forming my flock, I purchased the best Hampshire or West Country Down ewes I could meet with. Some of them I obtained from the late Mr. G. Budd, Mr. William Pain, Mr. Digweed, and other eminent breeders, giving 40s. when ordinary ewes were making 33s., and using the best rams I could get of the same kind until the Oxford Show of the Royal Agricultural Society. On examining the different breeds exhibited there, I found the Cotswolds were beautiful in form and of great size; and, on making inquiries as to how they were brought to such perfection, I was informed that a Leicester ram was coupled to some of the largest Cotswold ewes, and the most robust of the produce was selected for use. The thought then struck me that my best plan would be to obtain a first-rate Sussex Down sheep to put to my larger Hampshire Down ewes, both being the Short-woolled breed. . . . With this object I wrote to Mr. Jonas Webb to send me one of his best sheep, and he sent me a shearling by his favourite sheep Babraham. I went down the next two years, and selected for myself; but the stock did not suit my taste so well as the one he sent me, and I did not use them. I then commissioned him to send me the sheep which obtained the first prize at Liverpool, and from these two sheep, the first and the last, by marking the lambs of each tribe as they fell, then coupling them together at the third and fourth generation, my present flock was made."

Mr. Humphrey found his first difficulty in loss of size, and to obviate it drafted out his finest and smallest bred ewes, replacing them with the largest Hampshire Down ewes he could find that suited his fancy, and on these he continued to use the most masculine and robust of his own bred rams. This policy entirely succeeded, and, as he himself said, "beyond what I could have expected."

Oak Ash is eight miles from Wantage, Berks, and was named from an ash tree which grew up within the hollow trunk of an ancient oak, but which was removed and replanted where it now stands. It is an estate of 600 acres all under tillage, and without water meadow. It is nearly all good, and rather strong land, and has been known to grow one load of marketable wheat per acre over the whole wheat area. The house is well placed and commodious. Mr. Humphrey, at the time under our notice, was the proprietor of this and other lands, and remained there until his death. He unquestionably possessed in a high degree the peculiar genius

required in a first improver of stock.   It is a faculty which
must be implanted by nature, and comes to few.   He is
described as a fine-looking man, a capital public speaker, a
keen man with the gun, and an excellent shot.   He was a kind
master and good neighbour.

The following is a statement made to me recently by Abraham Hopkins, who lived as shepherd with Mr. Humphrey
from 1842 to 1868, and, therefore, from the date at which the
Babraham Southdowns were first used down to Mr. Humphrey's death: — "When Mr. Webb's sheep came master
would stand and look at him for two or three hours; or when
a good lamb fell from a favourite ewe, he would stand and
look at it, and move it about, for an hour or more.

"He took first prize with a West Country Down ewe at
Oxford in 1840.   It was not, however, till 1842 that he hired
his first sheep from Mr. Jonas Webb, and he had, in all,
three sheep from Babraham, for which he paid 60 gs. each
for the hire.   He had them at intervals of about two years,
and these were all the rams he ever bought or hired from Mr.
Jonas Webb or anyone else—one of them was named Thick-
thorn—but, with these exceptions, he used his own rams all
the time.   The ewes were drawn to these rams with the
greatest possible care.

"He only once bought ewes.   They were bought in a lot of
100, and of these Mr. Humphrey had twenty-five, Mr.
Rawlence twenty-five, and a neighbour fifty.   The ewes were
picked one by one, and Mr. Humphrey had the first pick.
The ewes were just outside the house," and Abraham Hopkins
says "he had noticed the best beforehand, and when he was told
to go in and pick one he went for her first.   She was a perfect sheep, and she bred John Bull, which beat All England,
and he was the father of Comet, which took first prize as a
shearling at Chester, and also at Warwick as a four-tooth.
Kettledrum was another son of John Bull, and took first prize
as a shearling at Leeds, and first at Battersea in 1862.

"Besides these ewes, no others were bought, unless it

might be one or two which struck Mr. Humphrey as desirable. One of such bought ewes bred Jack Tar by a ram from John Bull's strain. Such fresh blood was used with great caution and never directly. Thus Jack Tar was given a few ewes, and their ewe lambs were saved as dams for rams. It was, therefore, only after being well mixed with the blood of the flock that new blood was allowed to permeate it.

"Every lamb was marked as it fell, and those which showed any breachiness or coarseness were notched at the top of the ear; and no matter how well these lambs turned out, they were castrated and went to the butcher. Every lamb in fact, which was not let for breeding was fattened off, and no ewes were ever sold. Only good ewes were kept for breeding, and all the rest were sold to butchers. The ewes which were thought good enough for the flock were bred from until they were worn out. One favourite was kept till she was fourteen years old, and her last lamb was Oliver Twist. This ewe had no udder for the last four years of her life, and Oliver Twist was given to another ewe. This ram was first in his class at Leeds, and at Battersea.

"In using sires Mr. Humphrey was very particular. Lambs were used cautiously, by giving each of the best about twenty ewes. If the stock proved satisfactory the ram was used again as a shearling, and in subsequent years, but if not he was sent to the butcher. He always kept back his best lambs from the annual hiring for his own use, and was not a buyer at other people's ram sales, neither did he ever introduce strange blood straight into his flock."

Mr. Humphrey died in 1868, and his flock was then dispersed. Mr. Canning had some ewe lambs, so had Mr. Parker and Mr. Budd. Mr. James Rawlence gave 60 gs. for a ram lamb, Mr. William King paid 50 gs. for one, Mr. Ferris 47 gs. for one, Mr. Child 40 gs. for one.

Mr. Rawlence never missed a year having a lamb or two, and Mr. E. Waters the same.

Mr. James Rawlence has already been mentioned as an

early breeder of Hampshire Downs. It is worthy of note that while Mr. Humphrey commenced upon a foundation of West Country Down ewes, which had, as already explained, originated in Southdown crosses made during many previous years, Mr. Rawlence's original flock was " of the Sussex breed." He commenced by drafting all the small and delicate ewes, and crossing the larger and stronger ones with Hampshire Down rams. Mr. Rawlence frequently used Mr. Humphrey's rams, and thus obtained a fresh mixture of the Hampshire Down and Babraham Southdown blood, which was introduced with great skill and caution. The flock was further refreshed by purchases of Hampshire Down ewes, to which he put his own rams and used their produce. Mr. Humphrey's rams were used on some of the best of his ewes, and they again furnished sires for his flock. This process of infusing new blood gradually, and of rigorous selection, at length resulted in a flock of the highest possible merit, and Mr. Rawlence, in consequence, is regarded by many as the father of the breed. No one has done more to fuse the various elements into one compact and typical breed of sheep, and the Bulbridge flock became at length the foundation of many others. Mr. Spooner, in 1859, speaks of Mr. Humphrey's flock as distinct from any others, and applies to them the expression *sui generis*, from which we infer that one more step was necessary before the Hampshire Down could be regarded as a uniform and homogeneous race. In the accomplishment of this object Mr. Rawlence took the leading part, and we may look on the Bulbridge flock as fairly representing the Hampshire Down as we see him at the present time. Among other breeders of the district which may be considered as the particular native home as the Hampshire Down, I would especially mention the late Mr. James Read, of Homington; Mr. Alfred Morrison, of Fonthill, who early brought his ample means and skill to bear upon the improvement of these sheep; Mr. Dibbin, of Bishopstone; the late Mr. Newton, of Dogdean; Mr. Parsons, of Micheldever; and Mr. R. Coles, of

Middleton Farm, Warminster. Other early breeders have already been mentioned, and others are, no doubt, equally worthy of notice.

The Improved Hampshire Down sheep is the heaviest of all the Down breeds, and is only excelled in weight by the Lincoln, and occasionally by the Cotswold, among the long-woolled races. Its extreme earliness of maturity is well known ; and although it has recently been contested that the Cotswold may be brought up to as great a weight, or even greater, by careful feeding from birth, the entire habit of the Hampshire sheep is more in favour of early maturity than any other largely-distributed breed. The fact that Hampshire ram lambs are habitually sold for service at seven and eight months old illustrates this fact. The rapid growth of the lambs is most striking to witness, and one pound increase per day is a record which could be surpassed any year in particular cases. The late Mr. Coleman, of the *Field*, pointed out that the Cotswold did not appear to advantage as a lamb, but with the Hampshire it is just the reverse, as he appears at the greatest perfection in July or August, when about seven to eight months old. Those who wish to see what these lambs can do would find a visit to Salisbury Fair in July interesting.

The Hampshire Down has been accused of carrying an ugly head, but this defect, however common in years gone by, is now remedied. The ram lamb can scarcely be too dark in feature for the tastes of buyers, but this must be accompanied with white wool. A dark tinge around the poll will consign a sheep to a low price at once, but dark features and a fair fleece might elevate the same sheep into the region of keen competition. The ears must be free from any mottled appearance, and should in summer be like a bat's wing. The shanks also should be of rich dark brown colour, and free from mottled appearance. Some distinguished breeders have held a position in spite of a certain lightness of tint, but no light-faced sheep finds favour around Salisbury, which is the capital of the breed.

The nose in the ram should be thick and bold, and the ewe
should carry a bold head of more feminine character. The
character and ampleness of this feature are seen at all ages,
and distinguish the breed from Southdowns in a marked
degree. The lips are black, as are also the nostrils, and the
eye is of rich yellow-brown and large or full. The ears are
long, and in the best types fall slightly outwards, giving the
idea of great width of poll. They are thin and mobile, and
are set forward when the animal is in an attitude of attention,
giving an idea of intelligence and liveliness. The ear of the
Hampshire is undoubtedly a character, and differs from the
shorter and rounder ear of the Southdown or of the Shropshire. The head is well covered with wool both between the
ears and on the cheek. The neck is of fair length, enabling
the sheep to stand with head erect, instead of being carried
horizontally as in the Leicester or the Southdown. It is thick
and muscular, and is considered to be a point of special
excellence and importance. The shoulder tops are wide, and
the girth behind the shoulders and of the entire fore-end must
be well marked to secure any attention either in the prize or
sale ring. The remaining carcase points are common to all
breeds, and it seems unnecessary in every case to insist upon
the importance of well-sprung ribs, wide loins, straight quarters, good legs, square and massive form, &c. These go
without saying, and are as important in the eyes of Hampshire Down breeders as in those of any other sheep masters.
The fleece is composed of exceedingly fine fibres, and is thick
on the skin, which is pink in colour. The slightly Roman
character of the face and the fine wool have no doubt partly
been derived from the old Wiltshire horned sheep, which
lies back in the pedigree. The quality of the flesh and the
colour have come through the Southdown, but the colour has
been deepened by selection. The length of ear has probably
been derived from an alliance with the Cotswold, made, as
already pointed out, by Mr. Twynam, and in this feature
there is a point of resemblance between this race and the

Oxford Downs. Let anyone who wishes clearly to see the peculiarities of the Hampshire study them in contrast to the other breeds named in these particulars, and he will have no difficulty in fixing in his mind the peculiar characters of the Hampshire.

Knowing the susceptibilities of breeders, it may be well here to state that such reference to characters derived from a mixed ancestry is no slur upon the breed as it at present exists. The Hampshire Down has been too long established as a breed, and too long bred entirely *inter se*, to be now charged with being of mixed origin. Every race of sheep already mentioned has been crossed, with the exception of the South-down and possibly of the Leicester. It, indeed, seems to be necessary, if robustness of type is to be maintained, to make as a first-step such crosses as were effected by Mr. Humphrey, Mr. Rawlence, and, as has been asserted, was made by Mr. Twynam with the Cotswold. One of the great arts of breeding appears to be judicious crossing, followed by continuous breeding and weeding.

The Hampshire Downs are well inured to life between hurdles. So accustomed are they to this method that when turned out they usually move about in mobs, often grazing in a semi-circle, the foremost being in the centre of the curve. The number of sheep which are maintained upon Wiltshire and Hampshire farms is extraordinary. We have, for example, on the College Farm at Downton, lambed down 550 ewes on 600 acres in a recent season, as well as having maintained 200 tegs. The summer stock, when the lambing season proved to be very favourable, has consisted of about 1,250 to 1,300 sheep and lambs, besides a dairy of thirty cows and young stock in proportion.

This almost rivals the stock which Arthur Young mentions as occupying Mr. Ellman's farm at Glynde a hundred years ago, but the greater weight of the Hampshire Down sheep must be taken into account. Such a stock can, of course, only be maintained upon the acreage named in seasons when food is abundant, and must in less fortunate circumstances be provided with hired keep off the farm.

The constant use of hurdles points to an artificial system of feeding, and there is no doubt that sheep-farming upon the Hampshire and Wiltshire hills is a much more complicated business than the same pursuit upon the hills of Scotland or Wales, or even of Yorkshire and Northumberland. The character of the land and of the climate of these southern counties favours a system of double cropping with fodder crops followed with roots, and this, when assisted with large importation of cake and corn, and the ability of the Hampshire sheep to stand close folding, is the secret of the large number of sheep maintained.

The latest development in the history of the Hampshire Down is the establishing of a society for the promotion of the breed and the regulation of a flock-book. A preliminary meeting, held during the Smithfield Club Show in 1889 in London, was largely attended by breeders from many counties, and the feeling seemed to be unanimously in favour of the formation of such a society. It was felt that this step was chiefly necessary in the interests of the foreign trade, and that the Hampshire men ought not to be behind other breeders in this matter. The large size of the flocks is the chief difficulty in recording pedigrees. What may be easy when 150 to 250 ewes are kept might prove troublesome on farms carrying from 500 to 1,000 ewes.

For crossing purposes the Hampshire is exceedingly useful. It was by the alliance of a Cotswold ram to Hampshire Down ewes, and also, I believe, by adopting the reverse course, that the foundation of the Oxford Down was laid by the late Mr. Druce, of Eynsham.

The simple cross between Cotswold and Hampshire is frequently made for producing wethers, and the result is an increase in quality of mutton and of lean flesh, as well as of wool. Every year large numbers of rams find their way into Lincolnshire and the midlands for crossing with long-woolled ewes. They form an excellent cross with Leicesters, Lincolns, and Cotswolds, and are often put to these ewes during their last year of breeding, for producing fat lambs or wethers.

The Hampshire breed is able to withstand severe climates. Mr. John Craster, of Craster Tower, Northumberland, has for several years past kept a flock of pure-bred Hampshires, and esteems them highly. His estate borders the sea on the bleak east coast of the most northerly part of England; and as these sheep are able to thrive and give satisfaction in such a climate they may be credited with a hardihood equal to that of any other English race of sheep.

LIBRARY OF THE UNIVERSITY OF CALIFORNIA

HAMPSHIRE DOWN PRIZE SHEEP.

CHAPTER VIII.

OTHER MIDDLE WOOLLED SHEEP.

THE OXFORDSHIRE DOWN.

As an offshoot of the Hampshire Down and Cotswold races, the Oxfordshire Down may well follow the description of the first-named parent. Recent as the Hampshire Down must be considered to be, the Oxfordshire followed at a still later date. Up to 1859 the classification of sheep at the Royal included only Leicesters, Southdowns, Longwools (not Leicesters), and Shortwools (not Southdowns). In 1860 the Shropshire sheep were awarded separate classes at Canterbury. At the same show attention was particularly directed to the excellence of the Hampshire and Oxford classes, but it was not till the Battersea meeting in 1862 that prizes were for the first time given to the Oxfords as a separate class. At the general meeting of the Society, held in December, 1861, it was announced that it had been determined to add to the prize sheet classes for Lincolns, Cotswolds, Hampshire Downs, Dorset, *Oxfordshire Downs*, Romney Marsh, Mountain, and Irish Long-woolled sheep. Accordingly, at the great meeting in Battersea Park which was held simultaneously with the second great International Exhibition held in London in 1862, the Royal Agricultural Society recognised the fact that Long-woolled sheep (not Leicesters) and Short-woolled sheep (not Southdowns) did not constitute exactly a fair classification of the breeds of sheep of Great Britain.

The progress of sheep breeding within the last forty years has been indeed extraordinary. That within the memory of middle-aged men such a classification as the above should have been thought satisfactory is in itself remarkable. A sorry show indeed would it be, and hard upon the unfortunate judges, were the massive Lincoln, the Cotswold, and the Romney Marsh once more mixed up together as Long-woolled sheep (not Leicesters); or Oxfords, Shropshires, and Hampshire Downs contested with each other for prizes. Happily this is no longer the case, and a disposition has been shown to recognise any well-established and distinct breed.

The Oxford Down is more clearly a half-bred or crossed race than any of the preceding. It dates its origin to 1833 or 1834, and was produced by a cross between the Cotswold ram and Hampshire Down ewe, effected by Mr. Samuel Druce, of Eynsham, Oxon. It is curious to note that Mr. Druce speaks of them as a cross between the Cotswold and Southdown in a letter to Mr. Pusey, but this is evidently only an indication of the want of clearness at that time in distinguishing between Southdowns and west country or Hampshire Down sheep. In a later communication to Mr. W. C. Spooner, Mr. Druce is more precise when he says: "The foundation of this sheep was begun about the year 1833, by using a well-made, neat Cotswold ram with Hampshire Down ewes." In 1833 the Hampshire sheep itself was scarcely what it is now, for that was seven years before the time when Mr. Humphrey thought of going to Babraham to purchase Southdowns for the improvement of his Hampshire flock. The Hampshire Down parent of the Oxfords we may picture as a somewhat loosely-made, big-headed, sour sheep, such as the Hampshire sheep were in the early days before they had tacked on the affix " Improved," and when Mr. Clare Read tells us they were " swarming at Illesley Fair " —not, however, exactly in the form in which we see them to-day. The stream was tapped higher up, so that probably neither the Hampshire nor Cotswold parent of the Oxford-shire Down were exactly as these breeds now appear. Mr. Druce, in 1853, called his Oxfords " half-breds," and Mr.

Clare Sewell Read, apparently in search of a name in 1854, suggested the appellation of "Down-Cotswolds." In 1859 they seem to have become fairly recognised as Oxfordshire Downs, and appeared under this designation at Battersea, as already noted. There appears to have been a good deal of intermingling of blood before Oxfordshire Down breeders settled into a line of their own. Mr. Druce early found that " good qualities can be better secured by employing the cross-bred animals on both sides than by using the first cross." Mr. C. S. Read tells us that "the (flock) owner formerly divided his flock into three parts, putting a half-bred ram to the ewes that were about right—a Cotswold to the small ones, and a Down to the coarser sheep." Gradually the breed emerged from this tentative condition, and probably, at least from the date at which Mr. Read wrote (1854), the breed has been kept distinct from either parent strain, and has been bred exclusively *inter se*.

As might have been expected, the flocks of various breeders indicated in their fleece and faces a preponderance either of the Long or the Short Woolled parentage, and to a certain extent this difference still exists, although it is disappearing fast. The speckled face which used at one time to distinguish the Oxfordshire sheep has given way to a uniform brown, although a splash of lighter grey is sometimes to be discerned. The finely cut profile and thinner nose, and the long fore-lock, together with the long and thin ear, are evidently vestiges of the Cotswold parent. So also is the looser coat that is to be seen on the Oxfordshire Downs. The dark face and the comparatively close fleece are derived from the Down, and the degree of character and uniformity which has been attained must be put down to careful selection for a period of about forty years. They were called " the glory of the county — the most profitable sheep to the producer, the butcher, and the consumer," in 1854, and after a whole generation of men have passed away they still hold their position.

It is scarcely a matter of wonder that the Oxfordshire breeders should have seen the necessity of founding a flock-book and society for the promotion of the breed. In the case of a comparatively recent race, points require to be established, and bad types must be discouraged, so as to form a fixed character as speedily as possible. This first point of importance has, we believe, been achieved through many generations of careful breeding, and the Oxfordshire Down is in all respects a distinct and thoroughly constituted race; and has added much to the agricultural wealth of this country.

It is curious to contrast the results which were thought satisfactory in the early days of Oxfordshire sheep, when Mr. Samuel Druce and Mr. Philip Pusey were in correspondence about them. The carcase weight of tegs, at from thirteen to fifteen months old, was judged to be in the case of these improved sheep 76 lbs., and that of Hampshires is put down at 68 lbs. I have less personal knowledge of Oxfordshire Downs than of Hampshire Downs, but suppose they have also greatly advanced since this estimate was recorded. I should say that cull lambs of the Hampshire breed from a good flock should weigh 85 lbs. to 95 lbs. at fourteen months old, and that good representative tegs should have accomplished 80 lbs. carcase weight at ten months without forcing. Wether lambs are generally to be seen at Britford Fair on August 12th, selling at 60s.; and I, in 1883, sold 100 on that day at 73s. each. I have no doubt a similar result could be obtained with Oxfordshire Down sheep, although the Hampshire is especially quick in maturing. Much has been said as to the diminution of sheep stock in this country, but it should be remembered that a vast improvement and advance have been made in the matter of rapid growth. What used to take thirteen to fifteen months is now done in eight or ten, and hence, although the flocks of the country are less numerous, they are heavier per head, and the amount of mutton delivered to the butcher is probably not less than it used to be.

There is a good demand for Oxfordshire sheep for America
and the British Colonies. The breed is also extending in this
country, and its classes are usually very numerously filled at
the great agricultural meetings. As the remark has been
made by a practical man that Oxfordshire and Shropshire
sheep are so alike that when classed together at earlier meetings of the Royal, " many considered them one kind of sheep,"
I will point out the characteristic difference which close inspection will reveal. The Oxfordshire head is longer, and the
profile is bolder and slightly more Roman and fine; the
Oxfordshire ear is long and thin, whereas the Shropshire has
a shorter and rounder ear. The Oxford carries himself a little
more gaily and sprightly, and his wool is rather longer and
looser than that of the Shropshire. The wool on the head of
the Oxford is longer, and more of the flowing nature of a fore-
lock. That of the Shropshire sheep is closer, fitting like a
continuous cap or helmet.

Shropshire Sheep.

Midland farmers are justified in the pride with which they
regard Shropshire sheep. These sheep are humorously said
by their admirers to be so thrifty that in looking for grass
they turn the stones over which lie scattered on the surface of
clover fields. This statement is not intended to be received as
a fact, but as one of those hyperbolic expressions which farmers love to employ. I once heard a gentleman descanting on
the merits of Shropshire sheep over two hurdles in the show-
yard of the Royal Agricultural Society. He was standing on
a lower bar of the hurdle, and was thus raised a little above his
audience. He said, " It is a farmer's sheep, a rent-paying
sheep, a tenant's sheep. It is a money-making, wool-produc-
ing, mutton-carrying sheep. It's a bank, a save-all, a frugal
living and quick fattening, hardy sheep." This was Mr.
Preece, of Shrewsbury, and he was full of his subject and in
entire sympathy with his hearers. We wish to give every

breed its due, and as far as possible to enter into the feelings of the various breeders. We entirely concur in every word Mr. Preece uttered, and regard the Shropshire as one of those valuable additions to the flocks of the country which have come forward during the last thirty or forty years. The Shropshire sheep was indeed the first breed which broke through that singular classification of " Short-woolled sheep which are not Southdowns," which used to be considered satisfactory in the earlier shows. In 1860, or perhaps earlier, they were awarded a class, and, two years later, the same privilege was extended to Oxfordshire and Hampshire Downs. They now form a principal feature at all the " Royal" meetings, and their breeders seem less afraid of the show-ring than those of any other class of sheep. As an example of this I will take Windsor, when there were of

Leicesters	41 entries.
Border Leicesters	31 „
Cotswolds	60 „
Lincolns	58 „
Oxfordshires	82 „
Shropshires	212 „
Southdowns	123 „
Hampshires ... .·.	78 „

As a rule the Shropshire breed heads the list in point of numbers at the Shows of the R.A.S.E. The breed has found enthusiastic supporters in and around its native county, and its dominion extends over Shropshire, Staffordshire, Worcestershire, Herefordshire, and into North Gloucestershire, while there are also flourishing flocks in Scotland and Ireland. It appears to have had a mixed origin, and the fusion of the various elements into the compact form of a modern Shropshire sheep has been a work of much ability and time. They were formerly spoken of as Shropshire Downs, but lately the term Down has been dropped; first, because Shropshire boasts no " downs" in the sense in which the term is used in Wiltshire, Hampshire, and other chalk districts; secondly, because the breed is scarcely of Down character.

It is difficult to imagine that the massive carcases, carrying a "leg at every corner," and covering so much ground, were derived in the first place from a diminutive breed described in 1792 as the Morfe Common sheep. These sheep were then considered to be a native race, black or brown or spotted faced, and carrying horns, the wethers weighing from 11 lbs. to 14 lbs. per quarter, and the ewes 9 lbs. to 11 lbs. after being fed with clover and turnips. The fleeces weighed 2 lbs. This appears to have been the parent form, and the work of improvement consisted in crossing with the Leicesters, Cotswolds, and the Southdown. These various crosses produced in the first instance a somewhat uncertain type, but as early as 1853 we find them commended in the following language in the report of the Gloucester meeting of the Royal Agricultural Society : —" The new class of Shropshire Downs was very successful, and it is to be hoped that the Society will recognise them as a distinct breed." They were at that time described as without horns, " with faces and legs of a grey or spotted colour; the neck thick, with excellent scrag ; the head well shaped, rather small than large, with ears well set on ; breast broad and deep ; back straight, with good carcase ; hind-quarters hardly so wide as the Southdown, and the legs clean with strong bone. They are very hardy, thrive well on moderate keep, and are readily prepared for market, tegs weighing on an average 80 lbs. to 100 lbs. each. The fleece is longer and more glossy than the other Shortwools, and weighs on an average 7 lbs." Thus the Shropshire sheep, as contrasted with the maternal ancestor which grazed upon the Longmynd Hills, had during sixty years doubled its carcase weight, and increased the weight of its fleece threefold ! Writing in 1858, Professor Tanner says : " Only a few years since any mention of the Shropshire Down sheep raised an enquiry, even among intellectual agriculturists, as to their character, and few, comparatively speaking, knew anything of them." The Staffordshire breeders, forty or fifty years ago, traced the descent of their Shropshires from the native Cannock Chase breed—a dark faced, long legged, slowly maturing sheep.

The Shropshire sheep seems to have been the result of a gradual evolution, brought about in the first place by crossing, but later by careful selection. Even recently it was considered difficult to breed them truly, but, by dint of care and perseverance, the speckled character of face has given way to a rich uniform brown, accompanied sometimes with a little grey or " mealy " colour about the muzzle. The head is well covered with a close-fitting cap, or helmet, of wool, extending well over the entire space between and in front of the ears, which are somewhat rounded and short. The neck is carried horizontally, and is very thick and rather short. The shoulders are neatly stowed, and the girth is very great.

We are not able to name any breeder who, like Bakewell, Ellman, or Humphrey, actually accomplished the bringing-out of the Shropshire sheep, as these men brought out the Leicester, the Southdown, or the Hampshire Down. The work appears to have been shared by many, and the progress seems to have been gradual. Looking back an entire generation, we must, however, give full credit to the great advances made by the late Mr. Henry Smith, of Sutton Maddock, near Shifnal, who was a successful breeder and prize-taker; the Messrs. Crane; Mr. Green, of Marlow; Mr. Horton, of Harnage Grange, Shrewsbury; Mr. Farmer, of Bridgnorth, whose ewes made as high as £15 each when his stock was dispersed in 1857, while none of them made less than £8. Mr. George Adney, of Harley, near Shrewsbury, also had a splendid flock, as had Mr. S. Meire, of Berrington, and Mr. G. M. Kettle, of Dallicott. These flocks, and others, must be looked upon as the foundation of the present breed, and their blood flows in the excellent flocks of the present day. The number of famous breeders of Shropshires is indeed extraordinary, and, to their credit be it said, by far the greater number are tenant-farmers.

The Shropshire sheep is known in almost every county of England, and is esteemed for crossing purposes, and especially for raising fat lambs and tegs. It is well-known

in Scotland, and is popular in Ireland. Its greatest rivals
are the two races most recently described in these pages,
the Oxfordshire and Hampshire Downs. The Oxfordshire
and Shropshire sheep are so much alike that it is not always
easy to distinguish them from each other; but the chief
points of difference are to be seen in the shorter ear and the
greater closeness of wool on the head of the Shropshire sheep.
This latter is no doubt partly due to the methods of training,
but there appears in the Oxfordshire sheep to be a resemblance in the matter of fore-lock with the flowing top knot
of the Cotswolds. The fleece of the Shropshire is also finer
and more down-like in character, while the Oxfordshire
Down fleece is often gathered into strands or locks after the
manner of the Long-woolled breeds.

Shropshire breeders have long kept individual pedigrees
of their sheep, and most of the good sires are named. Thus
we see sire Jupiter 3,560, dam by Touchstone 1,775. There
is a regular Flock-book kept and every sire can be traced.

The Shropshire Sheep Society has been longer in existence
than any other, and no pedigrees of sheep are more carefully
recorded. It is the first breed we have met with in which the
sires are distinguished by a Flock-book number. We have
little doubt that a good deal of the success which Shropshire
sheep breeders have met with in the foreign and colonial
trades is due to the careful manner in which the pedigrees are
worked out.

We are passing through an important epoch in sheep
breeding. Never was there a greater interest shown in this
branch of pastoral life, as indicated by the formation of
society after society for the promotion of various breeds.
The Shropshire breeders were followed by the Suffolk Down
and the Oxfordshire Down men, and the breeders of Wensleydale Longwools, and still more recently we have seen the
Hampshire flockmasters combine in an association founded
upon similar lines. Even the long-established and aristocratic
Southdown, despite its blue blood and indisputable lineage,

has yielded to the spirit of the times, and a society and a club have been formed. There are also societies for Lincolns, Cheviots, Cotswolds, Dorset Horns, and Improved Leicesters.

The Suffolk Down.

This excellent breed is one of the few survivors of the old county breeds of Down sheep which inhabited the Chalk Hills of Southern England, and extended from Norfolk and Suffolk through Essex, Kent, Sussex, Surrey, Hants, Bucks, Berks, Wilts, Dorset and other counties possessing chalk downs. Of these Down breeds we can only name three which are of importance, namely, the Southdown, Hampshire Down and Suffolk Down, and these last have been greatly improved by the Southdown. The original Suffolk sheep existed in famous flocks during Arthur Young's time in peculiar form in the neighbourhood of Bury St. Edmunds. Mr. Edward Prentice, the able Secretary of the Suffolk Sheep Society, has contributed an excellent account of the "genesis" of the breed, and a scale of points is also to be found in the sixth volume of the Flock Book. The late Mr. George Dobito, of Lydgate, was one of the earliest and most enthusiastic supporters of the breed, and the establishment of the Suffolk sheep as a distinct breed was mainly due to his efforts. While the breeders of Suffolk sheep willingly allow the improvement effected by the use of Southdown sires, and, it may be, Southdown dams, they maintain that existing flocks date back to 1810, and have been bred with " rigid adherence to purity of blood." The Suffolk sheep may be described as follows, adhering to the points laid down as necessary by the Suffolk Sheep Society :—Head hornless ; face black and long, and muzzle reasonably fine, especially in ewes (a small quantity of clear white wool on the forehead not objected to) ; ears a medium length, black and of fine texture ; eyes bright and full ; neck moderately long and well set ; shoulder broad and oblique ; chest deep and wide ; back and loin long, level, and well covered ; tail broad and well set up ; ribs long and well sprung, with a full flank; legs and feet straight and black, with fine and flat bone; woolled to knees and hocks, clean below; fore legs set well apart; hind legs well filled with mutton; belly well covered with wool; fleece moderately short, close fine fibre without tendency to mat or felt together, and not shading off into dark wool or hair; the skin is fine, soft and pink.

In examining Suffolk sheep the observer is struck with the blackness of the face and the general absence of wool upon the head or between the ears, which is so characteristic of Hampshire Down sheep. This is the principal characteristic difference in appearance between the two breeds. The difference between the modern Suffolk and the hardy and small horned Norfolk and Suffolk sheep, of which they are the lineal descendants, is another instance of the wonderful progress which has been made during the present century in the improvement of our live stock.

THE DORSET HORN.

The horned sheep of Dorsetshire form a singularly well-marked race. In all the other Down sheep we find short wool associated with brown faces and legs, but in the Dorset we see a survival of a white-faced, horned, and short-woolled race, which has been for a very long period associated with the chalk hills of the county. Description of the old Wiltshire croocks in some respects favour the idea that the present Dorset sheep are similar to those extinct Wiltshires, which disappeared early in the present century, owing to the westward progress of the Southdowns. The Dorset breeders appear to have settled upon the process of selection rather than of crossing. They possess a sheep with special characteristics, and they wisely stick to them. Crossing was no doubt attempted, both with Devonshire Knots and the Leicesters; but William Youatt informs us that the attempt was not successful, and the breed has been handed down as truly representative of the old stock.

The peculiarity which stamps the Dorset sheep is the extreme earliness of the time in which they bring forth their young. They take the ram in April and lamb in September, thus producing lambs fit for the table in December, when lamb is a luxury, and therefore commands a highly remunerative price. Fifty years ago the Dorset sheep was described by Youatt as entirely white, the face long and broad, with a tuft of wool on the forehead, the shoulders low but broad, the back straight, the chest deep, the loins broad, the legs rather beyond a moderate length, and the bone small. The ram carries a pair of finely-turned horns, and the ewe carries a crooked horn, but without convolutions or spiral turnings. They are described as hardy and good folders, yielding well-flavoured mutton, and as averaging when three years old from 16 to 20 lbs. per quarter of marketable carcase.

The marked improvement which has been effected since Youatt wrote is shown by the fact that at ten to sixteen months old wethers are now brought out at the same weights mentioned by him as fairly representing that of a three-year-old sheep. I am indebted to Mr. Thomas Chick, of Stratton, Dorchester, for some interesting facts relating to the Dorsets of the present day, in a letter dated February 15th, 1890, in which he informs me that Mr. John Kidner's first-prize wethers at the last Smithfield Show weighed above 5 cwt. 3 qrs. live weight, or a little under 2 cwt. each, and Mr. Samuel Kidner's second-prize pen scaled somewhat more. The fleece is also now much heavier. Fifty years ago it was computed as averaging 3¾ lbs., whereas now a breeding flock of ewes will clip 5 lbs. to 6 lbs. of wool; two-tooth ewes, 6 lbs. to 7 lbs.; and Chilver lambs nearly 3 lbs. each. Rams will clip from 8 lbs. to 12 lbs. each; all these weights being taken after usual pool washing.

Mr. Chick kindly replies as follows to several enquiries which I made, omitting particulars with respect to the weights of carcase already given :—

1. What was the appearance of the early, unimproved type?

They were small and light in their fore-quarters, with black noses, and horns curving upwards and backwards, but were always remarkable as good mothers.

2. Who improved them?

The breeders in West Dorset were the first to improve the race; one of the earliest was the late Richard Seymour, of Bradpole, who had, perhaps, the best flock of Dorsets from 1830 to 1840. Others followed his example, amongst them the late Matthew Paull, of Burstock and Compton Paunce-foot; John Pope and John Pitfield, of Symondsbury; the two Davys, of Horn Park and Netherbury; William Way, of Bradpole; Thomas Chick, of Eggardon and Stratton. These all helped much in their day in bringing the breed up to its present standard. This work has been continued by their sons and successors, and others who have taken up the breed. Notably in the latter class may be named Messrs. Henry Mayo, of Coker's Frome, Herbert Farthing, John Kidner, Samuel Kidner, and W. T. Culverwell, in the county of Somerset, who deserve great credit for bringing the breed more prominently before the public.

3. Were they crossed with any other sheep?

The improvement was effected entirely by selection, not by crossing with any other breed of sheep.

4. What are their points of excellence?

In a general way, the usual good points in all sheep. Their horns should not rise from (? above) the head, but droop slightly as they grow outwards from the head, and curl round neither too wide from, nor too close to the face. They should always have a nice lock of close wool on their foreheads.

5. What are their properties?

Their excellence as mothers and breeders of early lambs, and general prolific qualities. They will take the ram at almost all seasons, and bring two and three lambs at a birth, when well fed. Their mutton is of good quality, and their wool realises about the same price as that of the Southdown and Hampshire.

6. Who are the principal breeders?

There are so many now who may be considered principal breeders, who have large flocks of really good sheep, about the same standard, that selecting a few names would be unfair to the others.

7. Do the wethers fatten kindly?

They do, but the larger number are sold fat as lambs.

8. What weight per quarter do wethers make at ten or sixteen months?

This depends upon the food and treatment. In an ordinary way, perhaps, from 16 lbs. to 20 lbs. per quarter.

9. When do the ewes lamb down?

Generally from November 20th to December 20th in each year.

11. What is about the value of lambs at Christmas or any time when generally sold?

Lambs are generally sold fat in March and April, weighing from 40 lbs. to 44 lbs. dressed carcase on an average; price per lb. according to the market.

12. What soils suit them?

Their natural home being the county of Dorset, the soils of which are the Kimmeridge clay, in Purbeck, the chalk near Dorchester, and in West Dorset the oolite, it appears that any of these soils will suit them. Dorset Horns have been kept in Purbeck for generations, and extend all through the south of Dorset into Somerset, where many large flocks are kept round Chard, Ilminster, and Taunton, as well as in the neighbourhood of Bridgwater. Breeding flocks are now kept in other counties, especially in the Isle of Wight. Ireland also has taken some of the breed during the past three or four years. Canada and the United States are also buyers of Dorset Horn sheep. A few good specimens were sent to Mr. Wm. Rolph, of Ontario; and Mr. E. F. Rowditch, of Mass., imported some choice sheep in 1887 and 1888. The largest consignment was in 1889, when Mr. T. S. Cooper, of Pa., U.S.A., imported 153 Dorset Horns, including all the first prize-winners of the

year. As the wonderful reproductive powers of this breed
become better known they will, without doubt, extend their
borders, not only in England, but wherever sheep for mutton
and lamb are kept. I would add that within my recollection
about half-a-dozen flocks of Dorset Horn sheep only were
kept, and at the present time they are as numerous as Downs,
if not more so. This is in the neighbourhood of Dorchester.*

---

* I am much indebted to Mr. Chick for so fully answering my questions,
thus giving readers the main facts as to the properties of Dorset sheep up to
the present time. There is one point upon which his information does not
bear out the statements of writers on this breed, as to the time at which house
lamb is available for the market. March or April seems somewhat late, as we
are told that Christmas lamb is chiefly obtained from Dorset ewes lambing
down in September. Youatt tells us that many farmers carried out the practice of house-feeding on a large scale. A building is usually set apart for the
purpose, divided into coops for lambs, according to their age. Every evening
the ewes are turned into their respective divisions of the lamb house, and each
mother speedily recognises her offspring. They remain together until the
following morning, when they are separated, and the ewes driven back to the
pasture. About a couple of hours after the mothers have been taken away the
ewes whose lambs have been sold are driven in and held until the lambs have
emptied their udders of milk, when they are driven to a separate enclosure.
At twelve o'clock the mothers are again brought, and remain with their lambs
an hour or two. At four o'clock the foster-mothers are again brought, and after
an hour's compulsory sojourn are again removed. At eight o'clock the mothers
return for the night. The greatest attention is paid to the cleanliness of the
place, and the lambs are supplied with good wheat straw for them to nibble,
and pieces of chalk to lick. Such was the practice in 1837. Those breeders
who lay themselves out for this special trade still manage to have lambs in
their flocks in October and November, and these are ready for sale at Christmas.

CHAPTER IX.

MOUNTAIN OR FOREST BREEDS.

THE breeds of sheep yet to notice are denizens of mountains, moors and fells. They roam over the free expanse, and add life to the scene. They are collectors of revenue, and that from the most unpromising situations, and in fact constitute the best means by which the mountain pastures can be converted into wealth. Whether on the sterile uplands of the Scotch Highlands, the more southern Lammermuirs, the border hills of Cheviot, the wild moors of Northumberland, Cumberland, Lancashire, Durham, or Yorkshire, the peak district of Derbyshire, or the picturesque localities of Exmoor and Dartmoor, these breeds of sheep are to be found. They are unconfined by hurdles or sheep-nets, and innocent of those artificial systems of feeding and forcing which are so important in the management of lowland flocks. What would be the fate of a pampered Lincoln, Romney Marsh, or even of an Oxfordshire or Hampshire Down, if turned out upon the higher slopes of Ben Lomond or the hills of Ochtertyre ? How would they stand the storms of Cheviot or of the braes of Ballochmyle ? Truly, each breed of sheep or of cattle is adapted for its own particular surroundings, and bold is the man who would advocate one particular race as suitable for all localities. We shall commence our study of the mountain breeds with the Black-faces of the Scotch Highlands, and afterwards notice the hardy race of the Cheviots, the Herd-

6

wicks of Westmoreland, the Lonks and Crag sheep of the
Yorkshire and Lancashire moors, and finally describe the forest
and mountain breeds of Wales and Devonshire.   The subject
is truly pastoral, and to appreciate it thoroughly the reader
should call to mind the grand beauty of the scenes in which
these sheep form no unimportant part of the picture.   So
thought the immortal Burns, who ever and anon in praising
the beauties of his country sang of "yowes" and "hog-
gies" :—

> The lee-lang nicht we watched the fauld,
>   Me and my faithfu' doggie ;
> We heard nought but the roaring linn,
>   Among the braes sae scroggie.

Or, again, in his famous song :—

> Ca' the yowes to the knowes,
> Ca' them where the heather grows,
> Ca' them where the burnie rowes,
> Ma bonnie dearie.

THE BLACK-FACED SHEEP.

Although best known in connection with the Highlands of
Scotland, it is doubtful whether this race is of English or
Scotch origin.   We know with some certainty that they first
obtained a footing in Perthshire and Dumbarton about 120
years ago, and it seems probable that they travelled north-
wards from Yorkshire or Northumberland across the border,
and then gradually displaced an older white breed celebrated
for the fineness of its wool.

David Low says:—"This breed may be supposed to have
found its way into Scotland by the mountains of the North of
England."   Youatt says that "it is a common belief in Scot-
land (1837) that the Black-faced sheep are of foreign origin."
There are many traditions as to the advent of the breed.
The Spanish Armada has had the credit of bringing them.
Dr. Walker is also quoted by Youatt as mentioning a tradi-
tion "that this breed was first planted on a farm in Ettrick
Forest by one of the Scottish kings.   The flock contained

5,000 sheep, kept for the use of the Royal household, and from that stock the whole of the Black-faced sheep are descended." This is, in some respects, a circumstantial account, but it lacks evidence of authenticity, as the name of the monarch and the authority for the tradition are wanting. Mr. David Archibald, in his valuable paper upon " The Black-faced Breed of Sheep," contributed to the Highland and Agricultural Society's Transactions (vol. xvi., 1884), indeed fixes this tradition as relating to James IV. of Scotland, and this he does on the authority of Hogg, the Ettrick shepherd. Another hypothesis, which at one time was currently received, was that the Black-faced sheep originated in a cross between goats and sheep. Many other views have been expressed of a contradictory character, and after carefully reviewing them all Mr. Archibald has come to the conclusion " that the origin of the breed is uncertain," and we must, therefore, take its existence as a fact, and leave the vexed question as to whether England or Scotland, or some other country, was the original home of the Black-faces as one which cannot now be answered. It is, indeed, a matter of regret that the origin of a sheep which only appears to have been established in the Highlands of Scotland for the comparatively short period of 120 years should be a subject of speculation and uncertainty, but such seems to be the case. That they forced their way through a considerable amount of opposition is evident, for in 1790 Dr. James Anderson wrote that " the coarse-woolled sheep (the Black-faces) have been debasing the breed (meaning the old breed then extant) under the name of improving it." Such complaints are not new to anyone who has studied the introduction of all the new and improved races, as has been already shown. We may, however, rest assured that the Black-faced breed possessed properties which fitted it for the exposed situations of Scotland, and it maintains its position better at the present time than it did only a few years ago. There appears to be no trace of the white and fine-woolled sheep, with hairy tails, which at one time occupied the ground now held by the Black-faced breed, and there does not seem to be any evidence that they were modified by crossing with the new race, as was the case with the old Wiltshire sheep when the Southdowns arose into eminence.

Descending from the regions of speculation and tradition, we find that Perthshire and Dumbartonshire were the first definite homes of the Black-faced breed. Argyleshire was either at the same or at a slightly later period colonised by the breed, and from these centres it spread rapidly. These sheep found an improver in the person of David Dun, of Kirkton, who has been spoken of as the Scotch Bakewell. In the Statistical Account of 1795, he is described as having the best stock of Black-faced ewes that are to be met with in Scotland. "They are," it is stated, "completely muir ewes, and yet they weigh 48 lbs. (22½ oz. to the pound)," or nearly 17 lbs. per quarter.

The principal breeders eighty years ago are stated by Mr. Archibald to have been Welsh (of Earlshaugh), Weir (of Priesthill), Gillespie (of Douglas Mill), Robertson (of Broom-lea), Kersie (of Glenbuck), and Foyer (of Knowehead), the grandfather of the present tenants. All these men were living about fifty years since. At a later date the breed was still further improved or maintained by Mr. Foyer (of Knowehead), Mr. Watson (of Nesbit), Mr. Watson (of Mitchell Hill,) Mr. Craig (of Craigdarroch), Mr. Dryfe (of Barr), Mr. Murray (of Eastside), Mr. Miligan (of Kirkhope), Mr. Sandilands (of Cummerhead), and Mr. M'Kersie (of Glenbuck).

It is always interesting to compare the results obtained in the past with those of the present. I therefore take an extract from Youatt, representing what was thought fifty years ago to represent fairly the capabilities of the Black-faced breed. Youatt says:—"They have mostly horns, more or less spirally formed, but the females are frequently without horns. The faces and legs are black, or, at least, mottled. They are covered with wool about the forehead and lower jaw, and the wool generally is somewhat open and long, and coarse and shaggy." To this I would add that the face is rather black and white than black, and that this interchange of colour ought to be distinct and not blurred. This is what the Highlander calls the " panting," which should be clear or well defined. In less improved specimens of the sheep the spine is liable to be traced by a line of hair (kemps) rather than true wool. "The weight," says Youatt, "of one of these sheep when fattened is from 16 to 20 lbs. per quarter, and the weight of the wool laid or unwashed is about 5 lbs., and that of a washed fleece 3 lbs. . . . The weight of one of these Black-faced sheep is about 16 lbs. per quarter." Turning to Mr. Archibald's article, written in 1884, or fifty years later :—" The staple of the wool has been increased in length from 4 or 5 to 8 or 10 inches, and it has been known occasionally to be 15 in. long. The weight of the carcase of eild ewes on good farms commonly averages from 15 lbs. to 16 lbs. per quarter, while in the best flocks they are sometimes equal to 20 lbs."

The points which are now looked for in a perfect animal are a thick, broad face, nostrils full, horns low set at the crown and turned backwards rather than forwards, and with a division or clear space between them. The colour of the face should be black and white, with the black predominating, moderately clear and bright. The carcase points need not be enumerated. A slight tuft of wool on the forehead in young sheep is generally thought to be an indication of good wool.

The management of Black-faced sheep is simple. The events of turning out the rams, yeaning, castrating, and weaning are all marked; but the chief point in management appears to be the selection of the grazing grounds. The flocks require the closest attention in winter. As another authority, writing in 1884, said, "For successful sheep-farming, a careful shepherd is the all-important functionary." His best qualification is to direct the sheep according to the nature of the soil and climate and the situation of the farm, in such a manner as to obtain the greatest quantity of safe and nutritious food at all seasons of the year." It is a common practice to take turnips in the lowlands during the winter. This system of turniping is found to encourage the growth and muscular development of young stock. Ewes in lamb are also sometimes allowed a supply of turnips, but, if they can be brought through without it, there is less danger of mortality at the lambing season. One of the special dangers to this breed, if placed on too nutritious diet, is the growth of the horns of the male lambs before birth, which often causes the loss of both ewes and offspring. Lambing is general in April, and the lambs are allowed to remain with their mothers till the middle of July.

Of late years the Black-faced breed has, to use Mr. Archibald's expression, had a sharp tussle with the Cheviots; but recent severe winters have been the means of confirming hill farmers in favour of the hardier race.

The distinct character of this race of sheep gives peculiar interest to the question of origin. It is assumed, as we have said, as the most probable view that the race was indigenous to England rather than to Scotland, and that it had crossed the Border by mountain sheep walks passing between and connecting Cumberland and Northumberland with Scotland. The vast extent of the moors of the north requires to be stated. Commencing in the Peak district of Derbyshire, they extend by way of Glossop, Ashton, Dewsbury, west of Bradford and Harrogate, to Whernside, and the Westmoreland mountains, through the moors around Stanhope-in-Weardale, across to Hexhamshire, Longtown, and Otterburn, to Cheviot; or around the eastern boundary of the Solway, into Dumfries, forming what is known geographically as the Pennine Chain. The Black-faced breed of sheep has been naturalised for ages upon their wide extents of moor and fell, and are to be seen in great numbers around Rothbury, in Northumberland. There could, therefore, be no difficulty in their passing freely from county to county until they reached first the Lowlands, which they appear to have done at an early period, and finally the Highlands of Scotland, beyond the Grampians, about one hundred and twenty years ago. Interest in such matters is comparatively recent, but the silence, or the speculations, of early writers as to the origin of these sheep are in themselves evidences of the antiquity of the breed. We sometimes find it difficult to clearly separate the long-woolled races of England, as no one can look at Devon long-wools, Wensleydale, Leicester, Lincoln, or Romney Marsh sheep without feeling that they have many points in common. Similarly the Down breeds have many points of common resemblance, and may be marked off as forming a second definite group. The Black-faces and Lonks form a third perfectly distinct type. The Cheviot sheep form another class, while the Herdwicks comprise a fifth group, having little in common with any yet named or described. How, within the four seas which bound our land, so many distinct types originated, or became naturalised, it is now impossible to say, but one of two theories may be reasonably adopted. First, that they have had distinct origin, from several wild species. Secondly, that they have been specialised by peculiarities of soil, of climate and of food, as well as by the fancy or the requirements of breeders, as in the case of the various descriptions of pigeons, and many other domesticated creatures. Both views may in fact be held; for we may reasonably believe that there have been some four or five original types of sheep which, by crossing and cultivation, have at length yielded the many breeds which we now possess. There is also the possibility of importation, by accident or by intention, of sheep which have become naturalised among us. Presents of animals have always been favourite tokens of respect among kings, and it is possible that our Black-faced breed was obtained in this way. This consideration gives colour to the statement which ascribes the introduction of the Black-faces into Ettrick forest to James IV.; but no one has ventured upon explaining how he became possessed of them.

CHAPTER X.

MOUNTAIN OR FOREST BREEDS (*continued*).

THE CHEVIOT BREED.

THE great Cheviot hill forms a striking object in Northumberland. It rises south of Flodden and Wooler, and overlooks the fertile lands which slope southward and eastward from its base. Cheviot is composed of trap rock, and is evidently of volcanic origin. The soil is of better quality than that of the mountain limestones and grits which form the basis of the heath country, on which the Black-faced sheep find their home. Cheviot is clothed with sweet, short herbage to the summit. It is early and late covered with snow, and its great altitude exposes it to severe storms during many months of the year. Here is the home of the Cheviot sheep, which have been bred there from time immemorial. The contrast between them and the heath sheep is complete. They are white-faced and hornless in both sexes. The body is long in comparison, which has given rise to the expression long and short sheep in speaking of the two rival races. The wool is fine and short, instead of coarse and lashy as in the heath sheep. They are, like most mountain breeds, disposed to be light in the forequarter. The fact of a well-defined breed occupying a limited area such as Cheviot, and bearing no special resemblance to any other breed of sheep, is a curious fact, and one upon which little light can be thrown. The most probable explanation is that the Cheviot breed is a survival of breeds of sheep once prevalent in different parts of Scotland, more or less resembling each other, and the Cheviot sheep of the present day. While these races have disappeared, the Cheviot has held his own, and not only so but has been improved and extended into many other localities, both north and south of his native hills. Towards the close of the last century, and up to comparatively recent times, the Cheviot breed was slowly displacing the Black-faces, but, as already mentioned, a reaction in favour of the hardier sheep has set in, owing to a series of severe winters.

Mr. David Archibald tells us that little attempt at improving Cheviot sheep was possible until the end of the border feuds and forays which disturbed the peace of the border land in the days of Johnnie Armstrong and other blackmailers and freebooters. It was not until about fifty years after the Act of Union, in 1707, that Mr. Robson, of Belford, worked out the improvement of the Cheviot sheep. In Douglas's " Survey of Roxburghshire," published in 1796, and in the " Farmers' Magazine " for 1803, some interesting particulars as to the early improvement of Cheviot sheep are given. In the first of these publications it is narrated that, " Mr. John Edmistoun, late of Mindrum, Mr. James Robson, then at Philhope, and Mr. Charles Kerr, then at Ricaltoun, went to Lincolnshire about the year 1756, and bought fourteen tups, with which they crossed their sheep with great success." The statements made in the " Farmers' Magazine " are equally clear. Giving " an account of the Northumberland breed of sheep, and the progressive improvements thereupon made," a contributor, signing himself, " A Northumberland Farmer," incidentally mentions Mr. Robson's selection of these rams, proving that there was a current opinion in the district to this effect. These Lincoln tups so improved Mr. Robson's stock as to give his sheep a decided superiority over those of his neighbours, and for many years, after making this cross, " he sold more tups than one-half of the hill farmers put together."

This then appears to have been the origin of the modern Cheviot. Sir John Sinclair described them in 1792 as a fine-woolled breed, and appears to have named them. In a foot-note to Mr. Archibald's article on Cheviot sheep, it is stated that on the establishment of the British Wool Society in 1791 by the late Sir John Sinclair and other noblemen and gentle-men, several delegates were appointed to visit the principal sheep districts of England and Scotland to examine the different breeds, and report upon their merits. During these investigations a breed was discovered on the borders of Eng-land and Scotland which Sir John considered well suited for being bred and reared in Highland districts. They were White-faced, and from their length were called "the Long sheep," in contradistinction to the Short or Black-faced breed. "*To these sheep Sir John gave the name of the Cheviot breed.*" From 1800 to 1860 the Cheviot sheep were more and more on the ascendant, and the Black-faces disappeared from nearly all the best farms in the south of Scotland, except in the mountainous districts of Ayrshire and Lanarkshire, and even in these regions their grazings were encroached upon. Since 1860, as already mentioned, the tide has again turned in favour of the hardy Black-faced sheep. On the lower and grassy slopes of the mountains the Cheviot sheep maintains his position; but on the higher and less accessible tracts, where heather takes the place of grass, the Black-faced breed is best. The points of a good Cheviot sheep are in many respects similar to those of other good breeds. Omitting those carcase points which may be supposed to be secured by all breeders, the coat should be good in quality, thick, and free from "kemp" hair, and fill the hand well. The head, while not too heavy, should be bold and broad, well set off by a bright dark eye, and erect ears of moderate length, covered with clean, hard, white hair. The nose is Roman in type, the skin around the mouth black; and both sexes are hornless. I will not enlarge upon the necessity of well-sprung ribs, broad loins, good legs of mutton, &c., but leave my readers to picture a hardy, up-standing, white-faced, hornless, and fine-woolled sheep, capable of standing the severe winters of Cheviot, but, nevertheless, one degree less hardy and less independent than his rival, which, it appears, is once more, as in the history of some nations, driving out his former conquerors. Upon this point, however, it would be interesting to hear what the promoters of Cheviot sheep have to say.

Herdwick Sheep.

Visitors to the English Lake country, even of the agricultural sort, do not make sheep their first study, being more likely to praise the mutton than the animals which yield it. Dotted over the high grounds of Skiddaw, Langdale Pikes, and Coniston Old Man, tenanting the beautiful grassy parks around Grasmere and Ambleside, Kendal and Windermere, may be seen a rather small but picturesque sheep. If the visitor is fortunate enough to see this beautiful district in the merry month of May, when the first blush of beauty has not been worn from the delicate lacelike foliage, he may also notice the little black-headed and black-legged lambs of the Herdwick breed keeping close to their well-woolled mothers. This race has kindled a wonderful enthusiasm in its favour among the dalesmen of its beautiful locality. The story goes that "forty small sheep managed to save themselves from the wreck of one of those Spanish galleons," which, after Admiral Francis Drake and Hawkins had broken that famous line, were driven by tempest along the rugged coasts of the West of England down into the cruel rocks of the Scottish coast to certain destruction. On the sandy Cumberland shore at Drigg these forty sheep saved themselves and "were claimed as jetsam and flotsam by the lord of the manor." Thus Spain became once more connected with the history of the sheep stock of our country. The episode is interesting and stirring, and, if true,* the Herdwick sheep may be looked upon as the most lasting memorial of one of the greatest events in English history—the destruction of the famous Armada in 1588. For 300 years they have held their own, and are likely to do so as being perfectly suited to their habitat.

I cannot do better than quote the excellent description of the Herdwick as given by Mr. James Bowstead, than whom no one knows the race better. " The essential points of a Herdwick may be briefly summed up as follows:—A heavy fleece of fairly strong wool disposed to be hairy on the top of the shoulder, growing well down to the knees and hocks, pole and belly well covered, a broad bushy tail, and a well-defined topping head broad, nose arched or Roman, nostrils and mouth wide, teeth broad and short, jaws deep, showing strength of constitution and determination, eye prominent and lively, and, in the male, defiant, ears white, fine, erect, and always moving, as has been said, like a butterfly's wing. . . . The colour or markings of the face and legs is very important. There should be no spots or speckles, nor any token of brown, as these are considered sure tokens of a cross. When the lambs are born their legs and heads should be perfectly black, with the exception of a little white on the tips of the ears, and per- haps a few white hairs round the feet. These white hairs gradually increase, so that at six months old one-third or half the ear will be hoar-frosted, and there will be distinct bands of the same round the feet, shading off to the black of the leg, and by this time also about an inch of the muzzle will have become frosted too. This change of colour goes on until some, at the age of three years, are perfectly white, whilst others remain a kind of steel grey. . . . Horns in the rams are

---

* Like many other traditions this may, however, be open to doubt, as it is also stated on authority that the Herdwick sheep were originally derived from a ship stranded early in the last century, which was lost on the Cumber- land coast. The sheep were saved and driven up the country, and purchased by some farmers who lived at Wardale Head in the neighbourhood of Kes- wick.—J. W.

desirable, but not essential, and undoubtedly add much to the appearance, but otherwise are not much valued. When present they should rise out well at the back of the head, be smooth, and well curled. White hoofs are much preferred." This careful description shows that in the Herdwick we have a distinct breed, readily distinguished, full of character, and not to be confused with any other. I quote one more sentence from Mr. Bowstead which, like those already given, indicates the real love of a Cumberland man for the breed of his district, and a disposition to clothe them with a romantic interest. "There are many yeomen in the dales of Cumberland and Westmoreland whose flocks have been handed down from father to son for generations without a blot or stain on their pedigrees, and he would be a degenerate son who would dare to try a cross." Bravo! In such hands the traditions of a breed may safely rest, and its qualities run small risk of deterioration or alteration.

At the Chester meeting of the "Royal" in 1893 many good specimens of this breed were exhibited. The second prize ram was nearly black in fleece. The swarthiness of the wool did not seem to constitute an objection in the eyes of the judges, and most of the rams carried dark wool on the shoulders and mane.

The Herdwicks possess properties which adapt them for a mountain life. They are said to select as a resting place those parts of the pasture in which they are not liable to be overblown by snow and entombed in wreaths. Also to scrape away the snow to a considerable depth with their feet in order to find the short and scant grass beneath.

The management of Herdwick sheep partakes of the general simplicity of sheep farming in mountainous districts. The animals are very independent, and able in a great measure to take care of themselves. The monotony of their lives (and of that of their shepherds) is broken by such leading events as lambing, castrating, washing, clipping, dipping, and selling at the fairs at Cockermouth, Penrith, and Kendal, and many find their way from these fairs into various parts of the country. While most English farmers are looking upon the lambing time as a thing of the past, the Cumberland and Westmoreland dalesmen are just about to welcome the little strangers. The rams are turned out from about November 20th to Christmas, so that lambing may be expected from April 17th to May 22nd.

THE LONK.

It is in vain that we search for notices of "the Lonk" breed in any of the older standard works on sheep. Youatt does not name the breed in his memorable treatise, which extends over 600 pages, exclusively devoted to sheep. Under the heading LANCASHIRE, he tells us that "the prevailing breed (1837), is what is here called the Woodland horned sheep, a variety of the heath or mountain sheep. . . They are found pure or with almost every variety of cross; but the principal crosses, and which are decided improvements, are, according to the nature of the country with the Leicester or the Southdown, and by means of which both the carcase and the wool are increased in weight and value." It is also noticeable that in describing the farming of Lancashire in 1849, Mr. William James Garnett (Prize Report, vol. x., R.A.S.J.) states decidedly that "there is no breed of sheep peculiar to the county." In Morton's "Cyclopædia" no mention is made of this breed. David Low seems to have been ignorant of its existence. While, as already said, the old writers do not name the Lonk, Youatt evidently had him in his mind when he penned the following passage: "In the West Riding of Yorkshire and on the borders of Lancashire, a breed of short-woolled sheep has existed from time immemorial. They are horned, with mottled faces and legs; some of them, however, are white faced. They are called the Penistone sheep, from the town situated between Sheffield and Huddersfield to which they are usually driven for sale. There is the same or a kindred race in Craven. It has been crossed more towards the south of the Riding with the Cheviot and the Leicester, and has been improved by both. Towards the north it has been oftener crossed with the Heath sheep, and then the legs and faces are black, or grey or spotted." This appears to fit with a remark found below, that " the Penistone breed is a shorter and thicker description of Lonk."

About thirty years ago a notice of these sheep, by the late H. H. Dixon, known as " The Druid," appeared in volume ii. 2nd series, of the Royal Agricultural Society's Journal, from which we make the following extract:—" The hill ranges of Yorkshire and Lancashire are believed to be the earliest home of the Lonks. We find them extending north from Clitheroe over the forest of Bowland towards Lancaster, east of Colne and Skipton as far as Keighley and Ben Rhydding, and south along the ' back-bone of England' by Pendle Hill, Burnley, Todmorden and Bacup, almost to Blackstone Edge. The PENISTONE breed, a shorter and thicker description of Lonk, there hold the hills. Saddleworth has also a large and plain sheep of its own with white face and legs, and coarse bone. The Saddleworth is a slower feeder than the common Lonk. Derbyshire also has Lonks on most of its hills and peaks, and its flock masters often go over to ' report progress' at the Craven show."

The Lonks are a distinct breed, and are most valued on low lying, damp and mossy land. They are like the Heath sheep, black and white faced, horned in both sexes, carrying a superior fleece of fine, moderately long wool, which is closer in texture and more springy and elastic than the wool of the Scotch Black faces. The breed is peculiar for resisting the effects of a damp soil. The Lonk is a larger, thicker made, and better woolled sheep than the Scotch Black-faced, and was well represented at the show of the Royal at Chester in 1893.

The Limestone or Crag Sheep.

In connection with the last-named sheep, there is another
breed which is stated to divide the rough and undulating
moorlands which constitute so large a proportion of East
Lancashire and West Yorkshire. Both of these breeds were
well represented at the Manchester meeting of the Royal
Agricultural Society. Like the Lonks, they do not appear in
any older descriptions of British sheep, but their existence is
indicated rather than asserted. Youatt tells us that ' towards
the borders of Westmoreland the Limestone breed of sheep
are found. They are natives of that part of the country, and
singularly confined to it. It is a horned breed, with white
faces and legs, depasturing on a rocky limestone land."

Here we have the nearest approach to a description of the
present Crag or Limestone sheep which we can find. I had
the opportunity of examining specimens bred by Mr. Rowland
Parker, of Moors End, Pourton, Westmoreland, at the Manchester Exhibition of the Royal Agricultural Society so long
ago as 1869, and was able to describe them from life in the
following words : " Both sexes horned, face and legs white,
wool firm, intermediate in length, and inclining to the character
of short rather than of long wool. Mr. Parker's flock clip on
an average 7lbs. each sheep. Mr. Parker rears and feeds
up his wether lambs entirely on the ' inland ' ground, and
raises them to from 18 to 22lbs. per quarter at twenty months
old. The ewe lambs are kept in the ' inland ' until they are
one year old, and then go to the ' common ' or high
ground from May to October. They are again brought down
to the inland in October and put to the ram. Mr. Parker
speaks to the prolific character of the females. Out of fifty-
four ewes thirteen produced three lambs each, while the entire
fifty-four brought up ninety-six to weaning time." The Crag
sheep are well adapted for the dry and high lying moors of the
mountain limestone, and are able to subsist almost without
water.

THE DARTMOOR SHEEP.

The Dartmoor sheep of to-day are a large, long-woolled variety rivalling in size the Cotswold, Lincoln or Romney Marsh breeds. They are the result of crossing the original Dartmoor sheep with Leicesters and Lincolns and do not give the idea of a forest or mountain race. They must be very different indeed from the "wild Dartmoor sheep" or "ugly old Dartmoors" of which Youatt speaks. At the Plymouth meeting of the Royal Agricultural Society there was a fine show of Dartmoor sheep which appeared about as like to the old-fashioned Dartmoor breed as a London alderman might be to an ancient Briton. Allowing for the influences of show-yard training, we can only now regard the Dartmoor as one of the heavy long-woolled, hornless and white-faced races of sheep, with such an amount of the old nature as suffices to enure him to the severe winters of his native home.

THE EXMOOR SHEEP.

As in the case of the Dartmoor breed of sheep, time has wrought great changes during the last fifty years. Then the sheep of both Dartmoor and Exmoor appear to have been similar to the Dorset Horns, although some were polled. They were small in the head and neck, small in bone everywhere; the carcase was narrow and flat-sided, and they weighed when fat from 9 to 12 lbs. per quarter. They were the material from which the celebrated Okehampton mutton was derived, and they carried a fleece of rather short middle-wool weighing from 3 to 4 lbs. of coarse and inferior quality. Even then the Leicesters were working wonders with the Devonshire aboriginal sheep, and the result is seen in the remarkable improvement and complete change in size and appearance which has taken place. If space permitted we could give instances of the wild nature and roaming habits of the original Exmoors, but sufficient has been said to show that they were a genuine forest or moorland breed before they were crossed

7

with the civilising Leicester. The modern Exmoor is much less than the Dartmoor, and is horned in both sexes, white-faced and covered with wool of the same character as that of the Leicester sheep. The close affinity with the Dorset mentioned by Youatt is no longer apparent, although activity and hardihood are still retained, together with a superior quality of mutton.

WELSH SHEEP.

Mr. W. Little, of Aberamon, near Aberdare, says of Welsh sheep:—"Most of the writers on British sheep dispose of those of Wales in a single sentence or thereabouts, as scarcely worthy of notice." This is scarcely fair, as Youatt devotes twelve closely-written pages to the sheep of Wales, taking each county *seriatim*. He, in fact, bestows more attention upon this subject than any other old writer.

As might be expected, the sheep of the Principality are divided into those adapted for the mountains and those more suitable for the valleys. It is on the high and picturesque parts of both North and South Wales that the native races are still found. The richer and lower lands are stocked with Cotswold, Shropshire, Oxford Down, Leicester, or with cross-bred sheep, in which all these breeds have been used.

GLAMORGANSHIRE SHEEP

It is impossible to give more than a general sketch of the sheep of South Wales. As a representative county we take Glamorganshire, bounded on the south by the Bristol Channel, and on the east, north and west by the counties of Monmouth, Brecknock and Caermarthen. Mr. Little, in writing upon the sheep of this district, informs us that the mountain land of Glamorganshire, though not rising to a great elevation, is bare, bleak and unsheltered. Its average value may be put down at 3s. to 3s. 6d. per acre per annum, whilst the enclosed patches around the homesteads may be valued at 7s. 6d. to 10s. per annum. The sheep are small, with few good points in shape and symmetry. They have white faces and legs, and, as a rule, are without horns, and they grow a short close wool, not wholly devoid of kemps or hairs. Some of them have brown or tawny legs and faces, and these are considered good points, denoting hardiness of constitution. The average weight of the mountain ewe, when fat, at four or five years old, is 28 lbs., or 7 lbs. per quarter, whilst the wether at four years old does well if he comes up to 40 lbs., or 10 lbs. per quarter. They are justly celebrated for the quality and flavour of their mutton, which attains perfection at four years old. Attempts at improvements are generally in the direction of crossing the native breed with Cheviot, Scotch Black-faced, and other rams, but unless the produce is much better kept during the winter than is usually the case, the cross-bred stock prove a comparative failure. It is found that the hill sheep of the district are most safely and surely improved by importations from Cardiganshire and North Wales.

Radnor Sheep.

In the county of Radnor, on the hills of Brecknock and in the western parts of Montgomery, and parts of Merioneth, remains a breed of the native dark-faced sheep of Wales, a hardy, active race, developed by good management and selection into animals of larger size than the ordinary mountain sheep, and carrying heavier fleeces. They have been improved by the introduction of a cross of Shropshire and of Leicester blood. The old breed was very small, and a great point with breeders was a very large tail, heavily woolled, and a quantity of coarse wool or hair about the breast. The best kind of Radnor sheep have black faces, but many are of a tan, grimy, or grey colour, and some, of unquestionable purity of strain, have faces partly white. The rams are horned, and the ewes should be hornless. They are short-legged, light in the fore-quarter, and, though slow feeders, yield mutton of excellent flavour. At three or four years old the wethers weigh 14 or 15 lbs. per quarter. The wool is of fine quality and weighs 4 or 5 lbs. the fleece. The ewes are prolific, and good nurses, and large numbers of them are sold into adjoining English counties to breed fat lambs by crossing with Shropshire, Leicester and Cotswold rams. The principal fairs are Kington, Knighton and Builth.

SHEEP OF NORTH WALES.

That indefatigable writer, the late Mr. John Algernon Clarke, informs us that there is no material difference between the mountain sheep of the northern and southern counties of the Principality. The best strains of Radnor sheep have already been described as black-faced, although this character is by no means universal among them. Mr. Morgan Evans thus describes the ordinary mountain sheep: " They are principally white-faced, but some have rusty brown faces, some speckled, and others grey. The males are horned, and the ewes generally hornless, though they sometimes have very short horns, and are occasionally found with horns equal in size to those of the rams. The poll is generally clean, but sometimes a tuft is found on the forehead of the ram. The head is small and carried well up, the neck is long and the poll high. The shoulders are low, the chest is narrow, the girth small, and the ribs flat. The rump is high, and the tail long. The average weight of ewes is about 7 lbs. per quarter. The wethers weigh at three years old 9 to 10 lbs. per quarter, and the mutton is famous for its delicacy. The average clip of wool is about 5 lbs. per fleece of fine quality, but in some districts it is mixed with long hairs (Kemps) about the neck and back."

These little sheep seem to be naturally adapted for the barren and high-lying tracts which they inhabit, and no successful effort appears to have been made to improve them. In 1865 Mr. H. H. Dixon reviewed the various attempts. The result of a cross with the Scotch Black faces was an increase both in wool and weight of carcase but it was not persisted in, as the wool became coarse, and the mutton rather yellow. The Cheviot cross increased the weight of both fleece and carcase, but produced a sheep too heavy for the mountain grazings. It was also found that the mutton tended to become light in colour. No pure Welsh leg of mutton should exceed 4½ lbs. weight, and larger ones are doubtful in their origin. In 1885 we find the same prejudice and the same verdict. Mr. Little, who has been previously quoted as an authority on Glamorganshire sheep, wrote at that date as follows : " I have myself been trying to improve the native mountain sheep for the last twelve years, but I am compelled to confess with very varying success. Many north country men have made persistent efforts to substitute Cheviots or Cheviot crosses for the Welsh sheep, but have been compelled to abandon them, after buying their experience somewhat dearly. Cheviot crosses have succeeded better on the more grassy hills of Brecknockshire, which rest on the old red sandstone, but they rarely succeed on hills lying on the coal measures."

There was a fair show of these sheep at the Royal Show at Chester, 1893. The rams were all handsomely horned. The ewes were hornless, and bare on the head. The noses were black and the faces and shanks were white. The faces were generally slightly rusty, or shaded with light yellow spots on ears, face, and also on the legs. One of the first prize ewes was peculiar for a rusty or light yellow face. The weights of these sheep appeared to be from 12 to 14 lbs. per quarter in the case of shearling ewes, and up to 16 lbs. per quarter in the case of two-shear and older ewes. These were unusually fine specimens selected for competition from the best flocks.

CLUN FOREST SHEEP.

The Clun Forest sheep was originally a white polled variety, from 12 to 14 lbs. the quarter, carrying a fleece of from 2½ to 3 lbs. Radnor Forest and Clun Forest form the boundary between Montgomery and Shropshire. This tract has for many years past been mostly under cultivation, and the improvement of its sheep has progressed with the development of its agriculture. The hardy Clun Forest ewes have been mated with Shropshire and Ryeland rams, and the result has been an improved type of sheep, which still bear the old name. Like all composite breeds not boasting a long lineage, the breed has scarcely yet advanced to a fixed or undeviating type. The colour of the face varies from fawn-coloured broken with white, to black or mottled. Both wool and meat are of excellent quality, and the race also is possessed of the valuable property of early maturity, far in advance of most farm breeds. Much, no doubt, remains to be done, but that the material exists cannot be denied.

CHAPTER XI.

APPARENT DIFFERENCES IN BREEDS.

THE minute differences between breeds of sheep receive but little attention. If we glance over a description of a race given by a breeder, we usually find an account which might do equally well for almost any other kind of sheep. We are probably told that the breed is remarkable for producing high-class wool and mutton, that it is singularly well covered on the back, well developed in the leg and shoulder, up-standing, hardy, and rent-paying. We may wonder why these points should be particularly mentioned, since they are not peculiar to any one breed, but are claimed for all. And yet the answer to this objection would doubtless be that they are the most important points, and that, if not present, the sheep would scarcely deserve a detailed notice. So recently as at the Plymouth meeting of the Royal in 1890, I asked one of the best judges and breeders of a well-known breed of sheep to tell me what were the actual points of difference between his breed and another which bore a striking outward resemblance to it. His reply was very characteristic. He said, "Nothing is easier—the difference is this. These sheep [his] possess all the excellences of the other breed, and a good many more besides." This is all the information I could get, but it is evident that if a promoter of the other breed had been asked the same question he might reasonably enough have given a similar answer.

The difference in feature, form, and fleece, habits and aptitudes, among breeds is an interesting study if nothing more. Judges, no doubt, attach weight to these matters when acting at shows, and do not give a prize to any sheep which is not distinctly a characteristic specimen of its breed. It is certain that a " Southdown man " cannot properly judge Hampshire sheep, or a " Hampshire Down man " Southdown sheep ; and hence we may conclude that it is somewhat unfortunate that any judge should be called upon to decide between animals of more than one breed. Single-handed judging is in my opinion a better arrangement, for when two men, one a well-known Shropshire breeder, for example, and the other an Oxford Down breeder, are judging these two breeds, the advantage of two heads is lost. The Oxford breeder would be disposed to follow his colleague in the Shropshire classes, and *vice versa*. If not, he will lean towards the particular type which he has always studied, and believes to represent perfection. The consequence may be that an animal is decorated which, however excellent, is not a characteristic sheep of his own particular breed. A student of live stock naturally asks himself what are the differences between two breeds which resemble each other so closely that the catalogue must be consulted before he can be certain as to where one class ends and the other begins. Many breeds are very similar, and others are chiefly to be recognised by the style of trimming, shaving, or colouring in which they appear in a showyard. Still there are, differences which reveal themselves to closer observation and which require to be pointed out. As an example of such difference let us take, in the first place

OXFORDS AND SHROPSHIRES.

Breeders might well say that these two breeds are distinctly different from each other, but a novice might find it exceedingly difficult to express what these differences are.

If the heads are closely observed it will be seen that the Shropshire sheep has a rounder profile, and is more completely clothed with wool towards his nose and mouth. His legs are also woolled down to his feet in great profusion—the general appearance of the head suggesting the Merino. The ears are still more characteristic, for they are short, thin, and rounded, and often light coloured at the tips. The Oxford ear is longer and thicker, and shows the origin of the breed from Hampshire and Cotswold parentage. The face of the Oxford is more wedge-shaped, the muzzle and lips are thicker, and the nostrils are more expanded. The face of the Oxford is more varied in colour, from rich brown to the same with splashes of light grey. The ear is the most striking point of difference, and the contrast between the large and bold ear of the Oxford and the rounded, short ear of the Shropshire is very marked. Many Shropshire sheep at Plymouth exhibited folds of skin about the throat and neck which forcibly reminded the observer of the Merino type. Loose skin about the throat such as many of those sheep carried would be considered objectionable by judges of Oxford or of Hampshire Downs. A Hampshire sheep thus furnished would at once be dubbed as " throaty," and would not command a high price in a ram sale. This feature was, however, apparently accepted as correct by the judges of Shropshires.

Leicesters and Lincolns.

The chief differences between Leicesters and Lincolns are seen in the larger and bolder heads of the latter, which are much more robust in type. The modern Leicester is also smaller in carcase and finer in bone than the Lincoln, which is paramount in point of size and weight of all our breeds of sheep. The Lincoln wool is displayed in large and bold masses, and is denser, stronger, and very much longer and heavier than that of the Leicester. The face of the Lincoln is more uniformly white, whereas dark or black spots are often to be seen on the ears and faces of the Leicester.

SOUTHDOWNS AND HAMPSHIRE DOWNS.

These two breeds are easily distinguished. In the Southdown is seen absolute perfection of form. We shall never see him surpassed in this particular by any breed. The plum-like outline, short and carefully-trimmed coat, and small amount of waste or offal are distinguishing characters. In size they are much less than the Hampshire Down, and, as is often the case with small animals, their symmetry is beautiful. The colour has become progressively lighter during the last thirty years, and in some of the specimens (not, however, decorated) the face might also be described as white, or very light grey. The head is dish-faced or flat in profile, and the ears are short and round, and often light in tint. The colour of the face of most of the prize-takers was a light fawn.

The Hampshire Down is much larger and bolder in form, and falls little behind the Southdown in fulness and symmetry. The old faults of neck, shoulder and rump have long disappeared under careful breeding, and for width of carcase and utility of form they will give way to no breed. Still, the Southdown must be considered as superior in its exquisitely rounded contour of form. The head of the Hampshire is almost black, and well covered between the ears, which are long, and fall away from the head, giving great width to the pole. This lopping of the ear may be carried too far, but must be considered as characteristic. The short ears of the Southdown are more erect, and are set rather more within the outline of the head. The nose of the Hampshire is thick and bold in the ram, and more rounded than in the Southdown. The Hampshire is cleaner under the throat than the Shropshire, as already mentioned. It is next to the Lincoln in actual weight. The chief point of excellence in the Hampshire Down is its extreme earliness of maturity. No breed can touch it in this particular. It is as a lamb that he is seen at perfection, whereas, with all respect to a recent result, Cotswold lambs cannot compare with them in this respect.

LINCOLNS AND WEST-COUNTRY LONGWOOLS.

The West-Country Longwools appear to have been based upon the original Devon Long-woolled sheep, improved by Leicester and Lincoln crosses. We were not able to see any particular distinction between this breed and Lincolns, except the artificial red colour, which did not add to their attractions. They are a useful breed of sheep, and well adapted for the West of England, and possibly for other localities also.

DARTMOOR AND EXMOOR SHEEP.

These two breeds are very distinct from each other. In the Dartmoor we have a large long-woolled, or, at least, heavy-fleeced sheep, rivalling in size the large, long-woolled races such as Cotswolds, Lincolns, or Romney Marsh. It is hornless and white-faced. The Exmoor is horned in both sexes, white-faced, covered well with dense wool of rather long staple, and massed together after the style of the Leicester.

It is much smaller than the Dartmoor sheep, and has more of the character of a mountaineer.

THE SOMERSET AND DORSET HORNS

constitute another very distinct breed, by no means resembling the last two. The white face, pink nose, horns and short wool, as well as the lighter form, closely mark out the Dorset horn as a definite race.

COTSWOLDS AND LEICESTERS.

These two races are easily distinguished. The Cotswold fleece is usually seen in bold, open curls, contrasting with the closely turned spiral curl of the Leicester wool or the dense massive wool of the Lincoln. The Cotswold also boasts a rather long and erectly-carried head, and long top-knot, whereas both Leicester and Lincoln have short and horizontal necks, and a tuft of wool rather than a flowing top-knot. The features of the Cotswold are finer in outline, narrower, and whiter.

CHAPTER XII.

THE MANAGEMENT OF SHEEP.

THE EWE FLOCK.

THE flock is always composed of ewes of various ages. Much has been said of late years upon the advantage of breeding from ewe lambs, but I cannot approve the suggestion. Nature will not be hurried with impunity, and for an animal to undergo the trials of maternity when it is not itself arrived at half its proper size is repugnant to her laws. If ample frames are to be perpetuated, the females should be fairly mature before they are placed in the breeding flock, and this is best accomplished by allowing them to bring their first lamb at two years old. This opinion may be challenged, but upon what ground but that of greed can the alternative be defended? Let anyone apply the rule to the case of other animals, and he will see that to begin breeding from females at too early an age is inconsistent with common sense. The case of sheep is peculiar in this respect, that there is no choice between lambing at one year and at two years old.

With other animals a compromise may be effected in terms of half-years, quarters, or months, and the different practices of breeders are thus only slightly divergent; but a whole year, or at least nine or ten months, is a long period in the life of a young ewe. I am, therefore, of opinion that as we must choose between half a year and a year and a half before mating our young ewes, we had better err on the side of leisure, and I think ewes thus treated will last longer. It should also be remembered that a few years ago ewes were often kept from breeding lamb until they were three years old, with very good result. We have now universally adopted two years as the best period, but the age is not likely to be further lowered with advantage.

The case of ram lambs stands upon a different footing altogether. They have not to undergo the protracted trials which, with the female, extend through the period of gestation and nursing, and, besides, ram lambs may be lightly worked if it is thought desirable, whereas there is no mitigation for the female. The best answer to these criticisms would probably be from the experience of those who think otherwise, but I should even then doubt whether a better result in the long run would not be obtained by those gentlemen who are thus working at a high pressure, if they adopted a less rapid course of action.

Our flock will then consist of one-fourth part, or rather more, of two-tooth ewes; one-fourth, or rather less, of four-tooth ewes; one-fourth of six-tooth ewes: and one-fourth of full-mouthed and over-year ewes. In a flock of 600 ewes the relative numbers would be about as follows:—

160 two-tooth ewes.
155 four-tooth ewes.
145 six-tooth ewes.
140 full-mouthed and older ewes.
———
600

This proportion allows for losses, and maintains a young flock at a maximum value.

Even by adopting this course it will be found that the two-tooth section requires the greatest amount of care both before and after lambing. They need better keep during pregnancy, and more liberal treatment after yeaning. This is done without much difficulty, as the extra indulgence consists either in giving a small allowance of cake or corn, or somewhat better hay and better fodder—or it may be cut hay—

while the older ewes are eating long hay, or hay and straw chaff.

The lambing period extends from January 1st to about the end of March. The well-being of a young lamb depends to a greater degree upon its food than upon the mildness of the weather, and these young creatures, if well supplied with milk and other foods, will stand any ordinary degree of cold and driving rain or sleet. Sheltered behind their mothers, or crouching beside a sheep trough, or behind a fence of thatched hurdles these young creatures will look bright and cheerful even in the severest weather—but we are anticipating.

Success in the lambing pen depends upon the previous treatment of the flock, and it is necessary to briefly sketch out the proper course to be pursued. The most trying months for early-lambing flocks are November and December. It is during these months that the change of food from grass and green food to turnips takes place. It is then that heavy falls of rain render the folds slippery and sodden. It is then that the farmer may delay too long the cutting of his hayricks. It is then that mistakes are made, and as these mistakes do not always tell disastrously, they are perpetuated. If a ewe flock is to be kept healthy it should as far as possible be kept on grass land. Those who enjoy the privilege of the grazing of a park have a capital opportunity; while those who must maintain their flocks entirely upon arable land are more liable to mischances. In any case, the rules for success are : —A small allowance of turnips and a full allowance of dry food. A firm and dry lair is of scarcely less importance, and I believe that, taking one year with another, the flockmaster who insures these conditions will experience what is called "luck" during lambing. Swedish turnips are not good for in-lamb ewes, *i.e.*, white and yellow turnips are to be pre-preferred. Second growth rape ought to be avoided, and mangel leaves are not good for a breeding flock. Mr. Robert Russell many years ago gave his opinion against ewes being allowed to run on stubbles, as he thought they found poisonous·

weeds there which injured them. I think this conclusion is scarcely borne out by the practice of others, and that the dry footing, and exercise obtained in such back-runs are good for ewes. It is not, as a rule, necessary to "cake" ewes before lambing, but as soon as yeaning takes place, or shortly afterwards, we can scarcely treat them too liberally.

Winter Management of Ewes.

Lambing-time no doubt brings the heaviest mortality, but the evil is contracted earlier. Months before, when the trials of lambing are scarcely thought of, and even before the ram is turned out, the seeds of mortality are sown. Why is the death-rate so abnormally high among ewes? Why do we lose a score of ewes to every cow that succumbs to the trials of calving? This is not only a matter of comparative numbers. It is true that we keep many more sheep than cows, but the hard fact remains that 3 per cent. of our ewes die, and a loss of 5 per cent. is regarded by many sheep farmers with calmness, if not complacency.

Such a loss among cow stock would be considered ruinous, and would be looked upon as a symptom of grave mismanagement. On the other hand, there are flocks in which the mortality is reduced within much narrower limits. How far we can always trust the rosy accounts which we hear after yeaning has passed, may be a matter open to doubt. Sheep farmers are not entirely free from a laudable wish to put their best foot foremost, and we may well discount some of these glowing reports of no "losses."

The word "loss" has a somewhat elastic meaning, as no sheep is absolutely lost if a part of its value has been recoverable. We, however, know flocks in which either from constitutional vigour, the nature of the soil, or management, very little loss is experienced, and we are inclined to think that this immunity arises from the observance of certain simple rules. Making every allowance, it is no doubt true that the amount of preventable mortality among breeding ewes is very serious, and our object is to enquire how far improved management can lower it.

In the case of ewes we find most deaths occur during and immediately after lambing, and that fatal cases are confined to about three months out of the twelve. Still, as already said, the evil is often done months before death occurs, and consequently continuous care is needed. We have known a heavy mortality among two-tooth ewes in January, which had been shorn as lambs eighteen months previously. They showed symptoms of having contracted a chill, as the flesh under the skin was red with congested blood, and we attributed the unhealthy condition of the animals to the removal of the wool, and exposure to cold and rain. Much damage may be done to a ewe flock by turning the animals out on cold nights immediately after shearing, and a yard with thatched hurdles or a well-sheltered paddock will be found serviceable at this period. Better still is it to put off shearing until the weather is warm, and the loss of the fleece can be borne without risk.

One day's starvation will tell injuriously upon ewes. They should not be allowed to become ravenously hungry or to be mistimed in feeding, as such mismanagement may easily result in losses at lambing time. Over driving and dogging have often been mentioned as injurious to ewes, as they no doubt are, and careful shepherds know how the injury remains after the cause has ceased to be apparent. A fruitful cause of loss among ewes is exposure to untoward conditions in November. The ewes are then getting heavy in lamb. They have been maintained on wholesome food and on a sound footing through summer and early autumn. During October they have scarcely felt the effects of the changing season, but in November the nights become cold, and frosty rimes cover the grass. Hoar frosts are proverbially followed by rain, and the flock is exposed to those steady downpours which are characteristic of the month.

The ewes show symptoms of having had a " rough " time,

8

and stand with their ears hanging and their backs up. The footing is rendered insecure and slippery, and this becomes a source of danger, especially upon sloping ground. If the ewes are upon turnips the food becomes dirty, and the whole appearance of both sheep and sheep-fold is one of desolation. The shepherd sees the danger and probably warns his master, and there is a rush to the hayricks in order to counteract the mischief. In many cases, however, the evil is already done, and the effect is seen either in slipping, or, subsequently, in the birth of dead lambs, and the loss of ewes in lambing.

These evils are easily preventable by the exercise of a little forethought. The cropping of a farm can be so contrived as to preserve a piece of old seeds or stubble as a back run. Ewes can be folded upon grass or stubble, and have their roots carted to them in limited quantity. Orders can be given to remove the flock from sideling ground of sticky character during wet weather, and to place them upon a dry, level field. Attention to these important matters of general management are most important, but are readily overlooked, simply because changes of weather often find the farmer unprepared, and the mischief is done before he has time to turn round and face it.

Flockmasters who have a range of unbroken downs or the run of a park, are placed at an advantage in comparison with those who are restricted to the limits of an ordinary farm. We know of few pleasanter sights than to see a flock spread over a wide extent of grass land or lying down on the dry and elastic turf. Such ewes are likely to have luck in lambing. Old seeds and old pieces of sainfoin may in a measure make up for these advantages, and it is often with the object of securing a healthy position in bad weather that such pieces are retained. The contrast between such a lair and a close fold upon turnips with the animals knee deep in mire is enough to reveal the reasons for good and bad " luck."

Longwoolled sheep require more room and greater freedom than the Down breeds. These last may be run more thickly upon the ground, and are less impatient of folding. In all cases the above considerations will make it evident that light soils of rather low average fertility are on the whole more suitable for carrying ewes than stronger and better soils. It is not merely a question of quality but of room, as the invariable tendency of good land is towards heavy stocking, and for this purpose dry sheep are more suitable than a breeding flock.

The feeding of ewes during November and up to lambing-time is a matter of first-class importance. Hay should be given in small quantities before October ends. It is the best safeguard against gorging with roots. The appetite must be satisfied, and if its edge is removed by a little sound hay or hay chaff, the ewes will partake of their roots moderately and quietly. There is no need to give cake or any other concentrated or expensive food until after yeaning. An outrun over down or dry pasture, a feed of hay, and a few roots are all that is needed. If asked as to quantity, we should say that if half a pound of hay per head is given from say October 20th, and increase up to 1lb. per head per day by December 1st, it should suffice.

Supposing the ewes to have a good run over down or old seeds, 12lb. to 15lb. of roots per head would be enough, or about a cartload (12cwt.) to every 100 ewes daily. In mild weather ewes will not eat inferior hay, but when the temperature falls below freezing they are less fastidious. Again in the spring ewes begin to be dainty as to the quality of their hay.

Two-tooth ewes require better food than old stock ewes, and are usually kept in a separate lot. A rather better quality of hay and a better choice of ground will be all that is requisite, and cake and corn are no more necessary at this time of year for the two-tooths than for the main flock.

Ewe tegs may be allowed a small allowance of concentrated food in order to maintain growth and size.

Ewes thus kept will cost about 4d. to 6d. per week, according to the allowance of hay.   When kept entirely upon "what they can pull," together with a few turnips, the first figure will represent the cost, and 1 lb. of hay may be reckoned to cost 2d. per week at 53s. 4d. per ton feeding value.

Ewes should be maintained in good healthy store condition up to lambing-time, and it is only after parturition is over, and the ewe and lamb are ready to go out of the pen, that the quality of the food need be improved.   It is then that we should advise an increasing allowance of cake, oats, and malt culms for the ewes, and peas and malt for the lambs; for nothing pays better for pushing quickly forward with plenty of good food than a nicely-bred lot of lambs.

The rules already given for the management of ewes in lamb are extremely simple, and yet are constantly being broken. They are as follows :—(1) a firm and dry lair ; (2) plenty of dry fodder ; (3) care, gentleness, and regularity.

With such simple rules to guide him it might well be asked why any person should transgress them ?   To this question we reply that the peculiarities of the season, and the requirements of the farm are constantly bringing pressure upon the farmer. He is from these causes subjected to a sort of temptation.

For example, a heavy root crop is in itself an inducement to both farmer and shepherd to feed the ewes liberally.   In a sense they *must* be consumed, must be got rid of, and hence the known principle of giving roots in sparing quantities is neglected.   Again, hay may be scarce, and the farmer, in his endeavours to make it last, delays cutting into his ricks.   He knows by experience that, when once cut into, a rick soon goes, and he therefore delays making a commencement.   In the meanwhile, the ewe flock suffers.   Farmers know the heavy expense of chaff-cutting and of purchasing food, and the vast importance of keeping a check upon outgoings ; and hence they may go beyond the limits of legitimate economy, and suffer later by a stroke of ill-luck at or before lambing time. Another example, showing how damage is done, in the face of knowledge, may be stated as follows :—The farmer is naturally anxious to get in his last sowings of wheat, and as he relies to a great extent upon his ewe flock to clear the ground of the root crop, he runs the risk, and pushes on the ewes, so as to make room for the ploughs. The probable consequence is a loss among the ewes and lambs later on. Even good farmers often do what they ought not to have done, and leave undone what they ought to have done—and this, not so much from want of knowledge, as because of the exigencies of the entire farm. A certain sacrifice, sometimes of keep, sometimes of corn, in the interest of the entire business, becomes necessary, and mischief is the result.

This mutual dependence of the various departments upon each other, and the wish to avoid expense, constitute disturbing elements, and prevent us from always doing the evidently best thing for our sheep. We cannot entirely avoid these disturbing elements, but a proper balancing of stock to area, and the exercise of forethought, will do much to mitigate them. It is not so much want of knowledge of rules as a right and ripe judgment that is wanted, and these qualities are best summed up in the word *experience.*

We hear a great deal as to the expense of ram-breeding, and one of the chief expenses consists in the sacrificing of the other interests of the farm to the sheep.

It is the " single eye " which leads to success, but when corn, and horned stock, and work have all to wait upon the flock, and everyone has to stand aside for the shepherd, our success as sheep-breeders is balanced by less success as corn-growers or cattle-graziers. This point requires to be brought out, because it is the duty of the master to see that the whole machinery of the farm is kept moving, and that all the wheels are oiled. This is the difficulty, and the reason of certain courses being at times followed which the master-mind knows full well are not the best for some particular section.

Another fact should also be stated in this connection. Seasons are so various that good luck sometimes follows bad treatment of a flock; and bad luck occasionally follows the best management. Such instances, which every frequenter of markets hears of, tend to make flock-masters run risks simply because no rule is to be always trusted, and mistakes do not invariably cause mischief. Ewes ought to be kept in medium condition. For eight or even nine months in the year they do not require cake or corn, and the time to spend money upon them is when they are nursing.

Improving a Flock.

Whenever a permanent and improving ewe flock is maintained, the principle of selection and of weeding must always be steadily kept in view. The draft ewes should comprise not only the older but the faulty sheep. Whatever the type— whether Leicester or Southdown, Cotswold or Hampshire— the shepherd's crook, like the pruning hook, must ever be at work, drawing out the weak and defective. However good the flock, the draft will contain some inferior specimens; and the theaves, or two-tooths, ought to put the fresh head to the flock. The annual progress of a flock is effected by the withdrawal of the weaker members and the importation of the newest and the best, and to this improvement there is practically no limit.

The master who perseveres in his work may see his flock steadily improve in value to a point of excellence and of actual money value of which he at one time scarcely dreamed. We will hardly say, in the language of a late statesman, "beyond the dreams of avarice," but certainly he would be a rash man who dared to fix the possible limits of value to which a flock of carefully-bred and long-established sheep might not reach. Rome was not built in a day, and a sheep flock of commanding merit will probably have survived the original founder many a year before the public compete for members of it as though they were buying pictures.

The determining reasons for drafting ewes are, in the first instance, age and such defects as broken mouths, rupture, bad udders, &c. The careful breeder will, however, look further. He takes the opportunity to weed out even younger ewes of defective feature, fleece, and form. He will not tolerate a weak neck, a mean head, a watery or loose fleece, flat ribs, drooping quarters, and shabby legs of mutton. So far as he can he weeds out such animals, and in a word he "raises his standard," and goes up another notch. It is here that genius shows itself, and it is at this point that the merits of each animal are discussed with the shepherd, who is as alive to the importance of selection as his master.

Similarly the young ewes about to enter the flock are canvassed over with the result already mentioned, that the flock loses its tail, and gets on a better head. This is the very essence of improvement, and the higher the flock rises in public estimation the more difficult is the achievement of further excellence.

It must, however, be remembered—and who is likely to forget?—that the male, as well as the female element, is of first importance. We are, therefore, in the next place confronted with the difficult task of

Selecting Rams.

Every year we hear of high prices given for rams, and it is not uncommon to read of 100 guineas being paid for the hire of a sheep for a few weeks. Such sums would appear small to an Australian flock-master, who would probably be able to cite cases in which hundreds of guineas had been given for a sheep. The prices paid are the best indication of the estimate placed by breeders upon first-rate sires. Here is the other arm of the service : a good and improving flock of ewes mated with rams likely to still further correct their weaknesses and produce ewe lambs which will again add to the reputation of the flock.

Rams must be bought with the strictest eye to personal merit. Success does not depend upon giving long prices,

but in securing good sheep. In mating we may follow the
advice of the Quaker with reference to matrimony : " Don't
marry for money, but go where money is." So in buying
rams, do not select merely on the ground of high price, but
go to the best flocks, and be prepared to back your opinion.
No doubt mistakes are made, but, on the other hand, the
best will always command bidders among rival breeders, and
the price will run up—there is no help for it. A breeder
will always find it expedient to be a buyer, but he should
also keep back some of his own sheep for service. Uniformity is a point which should not be disregarded, and this
is scarcely to be achieved without consolidating the particular characters of a flock by breeding from and with itself.
Besides, by keeping back a few of the best lambs of a good
flock probably better sires are secured than could be obtained
elsewhere, except at extreme prices, and hence the very best
lamb or lambs should be retained, or let for a limited time,
as in all respects likely to improve and amalgamate the
elements of the flock.

Most flocks have a character of their own. It is the im-
press of the mind of the breeder reduced to palpable form
through his selective powers. The breeder carries his type
of sheep about with him in his mind's eye, and nothing quite
pleases him which does not come up to this image. His *beau
ideal* must have a well-covered head, a good " scrag," a wide
shoulder. He will not forgive light fore-quarters, defective
wool or colour, and hence this man's flock will probably
reflect his fancy more than his aversions. On the same
principle, if he is a judge at a show he will recommend no
sheep for a prize unless it is up to his particular standard,
and hence exhibitors look out if they have an inkling that he
is coming, and draw and alter their exhibits accordingly.
There is also the influence of soil, which no doubt affects
sheep, making them looser or tighter in the coat, larger or
smaller in frame, or lighter or darker in feature. These two
chief factors, the master's ideal and the situation, both act

upon a flock and give it a character of its own. It is desirable that if the character is good it should be extended and rendered uniform throughout, and it is this approach to absolute uniformity which is the great ambition of a high-class breeder. No one who has not worked in a flock, and with shepherds, can imagine the numerous types of sheep. Every feature is noted, even such matters as the colour of the eye, the bend of the nose, the set of the ear. With fond affection and keen interest the shepherd shows you a pair of lambs which match each other, just as an accomplished horse-dealer will, after immense trouble, secure a " pair " of horses —not two horses which will run together, but a real matching pair. How absolutely true was the great Lord Althorpe's remark, that no man is fit to be a sheep breeder who cannot sit for an hour upon a hurdle looking at one sheep.

I lay further stress upon this point because outsiders have no idea of the ability, the time and the patience which are necessary to make a judge of sheep. I am often struck with the points noticed by first-class judges. They are not seen by beginners till pointed out. Some of my readers may have indulged in a social pastime which consists in placing some fairly conspicuous article on a well-furnished chimney-piece or table, such as a thimble or ring. One of the party who has been banished from the room while the placing of the said article is being effected, is then admitted and asked to find it, and it is truly amazing to see him pass and repass it apparently with his eyes upon it, and yet totally oblivious to the fact. He becomes frantic in his efforts to discover it, and at last it suddenly strikes him, and he seizes it amid the laughter and keen enjoyment of the company. So it is with points of sheep. The fault is glaring enough when pointed out, but till it is pointed out the unskilled man overlooks it again and again. He is at a loss to know why 60 gs., 80 gs., or even 100 gs. are being bid for one sheep, while the next is knocked down at 6 gs., 8 gs., or 10 gs. If he happens to be sitting next a judge of sheep, and confides to him his difficulty, his mind is set at rest by a quietly-spoken word drawing his attention to the head, the neck, the hocks, the pasterns, the mouth, the ears, &c.; and when once mentioned there falls from his eyes as it were scales, and he sees—and not only sees, but wonders why he had not seen for himself. This is one of the reasons why it is necessary to study sheep in the fold and the pen, and not only by means of books and newspapers.

After rams have been bought and the flock has been weeded, we next come to the

DRAWING EWES TO THE VARIOUS RAMS.

In an ordinary flock rams may be turned in among three or four hundred ewes, but in high-bred flocks the ewes must be drawn into little lots or even individually put to sheep likely to suit them. Practically, small parties of ewes, each accompanied with a ram, will be found sufficiently minute as a system. The principle of selection is simple, but the judgment required is very considerable. Strong ewes will be best mated with a ram of finer quality, and fine ewes may be put with a stronger description of sheep. Light colour may be corrected with a darker shade and *vice versâ*. Weak fore-ends, weak necks, want of width over the ribs or loins may be met with exuberance in these particulars in the male, and thus the flock may advance still one step further in its development and approach to perfection.

BEST CONDITION FOR EWES.

Ewes ought to be brought into good condition before tupping by means of suitable food. A short lambing season is secured by a quick seasoning of the ewes, and this is promoted by liberal and stimulating feeding both before and at the time that the rams are out. Early turnips, good rape or kale, and in some cases about one quarter pound of cake help to bring ewes on, and such good and therefore expensive feeding may be discontinued as soon as the ewes have conceived. An old notion among shepherds is that a heavy fall of lambs is best secured when the ewes are "fresh" at the time they go to ram, and afterwards are put on poor dry keep. Fat ewes (if not overfed) are always likely to produce early and abundantly. Our own system is after the ewes have been served to put them together in one large lot, placing one ram with them to take those which "turn." That the number of twins depends in a great measure upon management is certain, as some breeders always secure a much larger crop of lambs than others, and the main secret is no doubt that the ewes should be in good condition while with the ram.

DIPPING.

Ewes should be dipped before they become heavy in lamb. July is a favourite month for the purpose, but it may also be done in September or October. The dipping of ewes has a twofold object—the destruction of sheep ticks and the promotion of a clean and healthy condition of the skin. Prepared dips are usually sold in packets ready for use, and in these days few concoctions are made at home for any of the old purposes of the farm. It is the tendency of the age in both domestic and outdoor occupations — the home-made recipes give way to the purchased article. So it is with sheep dips. The mixing up of white arsenic, pearl ashes, sulphur and soft soap is relinquished in favour of a purchased dip with directions for use printed on the wrapper, the material to be at once mixed with a prescribed amount of water. The best time for dipping is about one month after shearing, if for no better reason than that less of the material is taken up than when the wool becomes longer. This applies more to Long-woolled sheep than to the Down breeds, which may be conveniently dipped after harvest.

DRAFTING EWES.

This is one of the most important events of the year in the history of a large breeding flock. Shearing ought to come first, and a few days should elapse to allow the ewes to get over their somewhat shabby appearance after they escape from the hands of the shearer. Ewes look somewhat wretched when first shorn. Less pains is taken with them than with market sheep, and they are the proper material for youngsters to try their 'prentice hands upon. A fortnight later the marks of the shears have disappeared, and the ewes are less likely to be seen with their backs up and their necks craned, as we have often noticed them on a cold day late in May. It is then that the master and his trusty shepherd may look the flock over with a view to removing old and faulty ewes.

AGE.

It is wise policy to keep a young flock. It is thus maintained at a maximum value, and the draft commands the highest price, and will sell either for breeding or fattening. The usual draft is, as a rule, young, for three crops of lambs only leaves the ewe four and a half years old, which is far within the allotted span of a sheep's life. In a previous chapter, when writing upon the late Mr. Humphrey's system of improving the Hampshire Down breed, I mentioned that this noted breeder never sold any draft ewes except to the butcher, and that he had been known to breed from a ewe until she was fourteen years old! It was at this advanced period of life that she bred Oliver Twist, the champion sheep of his year. Such a fact may well be laid to heart by breeders who consider one or two years over age as about as far as they like to go in keeping on ewes. A good ewe ought certainly to be kept as long as she will breed, and seven or eight years is not too old for this purpose. We should probably all be the better of remembering this fact, and not allow splendid ewes to go out of the flock at the comparatively early age of five or six years.

MOUTHING.

One of the first preliminaries is to "mouth" the flock by running them through a narrow passage between hurdles, and examining each ewe as she passes. Broken-mouthed ewes should not be kept unless for special reasons, but as long as a sheep has a good set of teeth she may be depended upon. A full-mouthed sheep is four years old. She is a two-tooth at fourteen months; a four-tooth at two years old; a six-tooth at three, and a full-mouthed sheep at four. At the usual time of drafting she will, therefore, have the full set of eight incisor teeth. At five years old her teeth may be a little wide, and at six she may have lost her corners, or the central and middle teeth may show signs of wear, and when this is the case it is time that she went along.

THE UDDER.

Just as it is important that a breeding ewe should be able to break her turnips and eat her hay, so is it of equal import-ance that she should properly support her lambs, and hence a healthy and perfect udder is of first-rate importance. Ewes should be turned in order to see that they are sound in the bag as well as on the tooth, and the inspection will probably decide the question as to whether she is to be maintained another season or to be sold off as a cull.

The udder is variously affected. Sometimes a teat is corded, that is, the canal through which the milk flows is obliterated and one or both halves of the bag is rendered use-less. In other cases the bag is found to be ruined by inflam-mation and suppuration, and to be unfit for its natural function. The ewes so affected are not fit to stand among a hundred good cull ewes, and should be fattened off for the butcher.

POINTS.

So far the ewes speak for themselves. Old, broken-mouthed, or, we may add, ruptured ewes must go, and a faulty udder is in most cases an unpardonable defect. It is less easy to still further cull the flock of its less desirable members. Great judgment is required in culling out the weak members, but the opportunity must be taken, and no doubt rigorous weeding is one of the secrets of improving a flock.

A weak, bare, or badly coloured head, speckled ears, when a uniform colour is the type, pink or badly coloured lips and nostrils, and spots where no spots should be, a rusty, sour, ugly head, in any breed, should be got rid of. It is no great matter if we cull beyond our usual draft, as there are plenty of opportunities in early autumn to replace, by buying a few good ewes.

Nothing looks better than good heads, and, strange as it may seem, a sheep's head, which is only worth 9d. at the butcher's, is worth a lot of money when carried on a good ram or ewe.

Next to the head and ears we look for good necks. Ewe-necked sheep never look well, and a good scrag is a strong point. Let us, therefore, as far as possible weed out long or hollow-necked ewes. A muscular neck indicates strength of constitution and good muscular development, and I have never known a sheep breeder who did not strongly object to a shabby neck. Mr. Ellman, the father of the Southdown breed, insisted on the importance of this point. Mr. James Rawlence, of Bulbridge, one of the oldest of our noted breeders, would not keep a weak-necked ewe, and no man who values his flock would buy a ram with this fault. The neck ought to be muscular, arched, tapering and neat.

Shoulders are as important as neck, and should be considered as follows: First, they must blend with the neck. They must be well laid back so as to produce thick "crops" and a great girth. Secondly, they must be wide over the tops.

Thirdly, they must be thick through the heart, from blade to blade. Nothing can be more effective than a good fore-end. If you try to think of it, imagine the sheep to be grazing with her head towards you, and you will then notice the grand effect of good shoulders. Deep floor to the chest and a prominent breast coming well forward between the forelegs completes this part of our picture. Next let us look at the ribs and back, the loins, the quarters, the let-down of the legs of mutton, and lastly at the general ampleness of form. There is no mistaking a good sheep, and when looking through a flock for drafting, every mean, under-sized, bad charactered, or defective ewe, must go. A good flock cannot be got up in a year, but each year tells. It is the object to take off the tail and put on a new and improved head to the flock every year, and thus to build up to the ideal which every good breeder carries in his mind's eye. This is drafting, or weeding, and no successful breeding can go on without it.

CHAPTER XIII.

THE LAMBING TIME.

As lambing time approaches, the flock-master begins to prepare for this important season. We know, as a fact, that lambs are often lost for want of care. Shelter for the ewes must be provided, and this leads us to the consideration of the lambing pen. There are two descriptions of enclosures for lambing ewes. One is the old-fashioned permanent pen, for which the rick-yard has often been employed. The advantages of this system are that the flock is near home, and that the rick-yard is a protected enclosure which, when well littered down and fenced with thatched hurdles, forms a very suitable place for the purpose. We have seen special walled enclosures, furnished with accommodation for the shepherd, and shedding for the ewes. The shedding is most conveniently divided into coops by means of hurdles, and in such a shed ewes will lamb safely and comfortably.

On large sheep farms this system is objectionable on account of the distance between the flock and their food. It is, therefore, the custom to make a pen near to where the ewes and lambs are to turn out after lambing. The position of the pen should have been fixed during the previous summer, and also the situation of certain hay and corn ricks determined. As threshing proceeds, the corn ricks yield straw ricks, which are made long, and placed so as to secure the greatest amount of shelter from the wind. A gentle slope towards the south is the best site, and in close proximity to a field of swedes, or of late turnips. The en-closure consists of a double row of hurdles, stuffed between with straw, and kept firm by means of a few posts and rails. About two feet from the outside wall, and on the inside, are driven six feet posts carrying a head rail or plate, and, rest-ing on this plate, and upon the outside hurdles, with a sufficient run or slope, thatched hurdles are fixed; thus forming a continuous narrow shed, which is again divided by hurdles into coops or cells. These coops are best open to the south and east, and backed to the north and west; and in such a position ewes and lambs lie warm even in the severest weather. Outside these cells, and inside the en-closure, the space is divided by hurdles into four or five good-sized yards, and a straw rick ought to occupy a central position with reference to the entire space. The shepherd's portable house is drawn up at a convenient distance, and with such a fold we may look forward to lambing with a feeling of confidence and security.

The forward ewes should be brought into the pen every night and lie upon straw. A good-sized heap of swedes should also have been provided, and hay racks or cribs should be placed around, so that the animals may receive a foddering when they come into shelter at about four o'clock in the afternoon. During the height of the yeaning, the shepherd remains night and day with his flock, and provided with a good lantern he makes periodical visits, carefully looking at every ewe. As soon as a lamb is born, it and its dam should be removed into one of the coops or cells, as already mentioned, there to remain for three or four days, until the lamb is able to follow its mother without difficulty, and until the two thoroughly know each other. When this is judged to be accomplished, the cell is vacated for other occupants, and the ewe and her lamb or lambs are transferred to one of the larger divisions of the pen.

As lambing proceeds, the various lots of ewes are classified and separated, as follows :—

9

1. A yard of ewes heavy in lamb.
2.   „   „   with single ewe lambs.
3.   „   „   with single ram lambs.
4.   „   „   with twins.
5.   „   „   and very young lambs.

The older lambs, with their dams, are, when from four to seven days old, allowed to go out upon the turnips, and it is interesting to watch these young creatures learning to fend for themselves, and imitating their mothers in their eating, choosing the softer parts of the turnips, nibbling at the rape or turnip greens, or sorting out the choicer portions of the hay.

Lambs ought to be provided with a corner for themselves at an early age. A few hurdles should be placed around so as to include some small troughs in which is placed a mixture of split peas, bruised oats, and finely-ground cake. Admittance is given to this enclosure by means of lamb hurdles, which, while allowing of the ingress of the lambs, is a bar to the larger-sized ewes.

The lambing pen having been constructed, and all arrangements made for the arrival of the lambs, we shall now give our attention to the important subject of parturition.

Sheep, although able to withstand the vicissitudes of our changeable climate, and to resist the injurious effects of a damp bed and night air, are nevertheless easily "upset." They are, so to speak, as hardy as the hares which play around them as long as they are well; but, for all that, no domestic animal gives way more suddenly and more hopelessly when attacked by disease.

A flock of ewes resting upon frozen ground, placidly chewing the cud, while the snow wreaths are forming around them, give the idea of extreme hardihood, and when the frost and snow are exchanged for mud and driving rain, the beholder may well feel astonished at the quiet indifference of these creatures to the most rapid changes of meteorological surroundings. Sheep are, however, liable to many diseases and sundry kinds of death, and when once a flock goes wrong the individuals composing it exhibit a weakness of constitution which the above picture would little lead a tyro to expect. All is well as long as the flock is kept healthy, but the mortality is terrible when the conditions of health have been outraged. These strokes of " bad luck" are familiar to most farmers, but may be avoided by strict observance of the laws of health.

The cause may be remote in point of time, but, when the blow falls, it usually falls heavily. At such times the entire flock has been known to succumb, or it may be that the loss is measured by scores of lives.

Constant forethought and observance of the rules of good management are the best safeguards, and when due precautions are always taken, the rate of mortality can be kept within bounds. A competent shepherd thinks carefully over every change of food or situation, and often, by a timely word of caution or practical suggestion, saves his master from loss; and, on the other hand, a careless or inexperienced man may in a few hours do incalculable mischief. Good general management, careful feeding, and judicious changes are better means of keeping a flock in health than any amount of tinkering or doctoring. A flock well treated during the months preceding the lambing season approach that critical period in robust health, with well kept fleeces, bright eyes, and clean faces. Their dung falls in the form of glossy black pellets, and there is no indication of scour or dirt about the tail. The animals, are, in fact, clean both before and behind, and all between looks thriving and lusty. When ewes come up to the lambing pen in the condition we have endeavoured to describe, we may hope for what is called good luck in lambing, and as we have no wish to describe a badly-managed flock, we shall assume that due regard having been paid to the previous management, our flock approaches the great event of the year in a healthy state.

Parturition is a thoroughly natural event. It is not strictly speaking an illness, but an act of exuberance. It is accompanied by evident uneasiness, and in some cases with great pain, but when once over the ewe seems to forget her troubles immediately, and promptly undertakes the cares of maternity. In normal cases—which, by the way, are by far the most numerous—the ewe should be carefully watched and attended to, but assistance ought *not* to be given. Patience should be exercised, and a somewhat long period of labour often ends in an easy birth. A watchful eye in such cases is all that is needed, and attention to the newly-born lamb. Its mouth must be cleared, and the first gasp for breath must not be stifled by the water and mucus which often envelop the mouth and nostrils. Exposure to intense cold or bitter weather might at this stage prove fatal, and hence the shepherd transfers his charges to a comfortable pen, and is content when he sees the youngster fairly struggled on to its long spindly legs, and groping about in its quest for the source of its first meal. This, then, is our answer to the question so often asked, "Should assistance be given to a ewe during lambing?" No interference, no inter-uterine examination, no forcing on the pains by pulling at the legs of the fœtus, but only carefully watching the course of events.

There are, however, times in which the shepherd needs to put forth his obstetric skill. This is necessary in all abnormal cases. Before a shepherd interferes in the process he should at all times think whether or not artificial help is needed, and if he comes to the conclusion that the ewe wants help he will give it, in the manner about to be described.

The ewe is laid gently upon her side, and the hand is carefully introduced into the *vagina*. If the *fœtus* is to be found coming forward in the natural position—*i.e.*, with his head resting upon his two fore-legs—immediate assistance is not required, and the ewe had better be released and allowed a little more time.

Examination may, however, reveal the fact that the lamb is presented in such a manner that assistance must be given.

The false or abnormal positions may be thus enumerated. They all have one common character—namely, variation from the normal position above indicated. These variations are either precisely defined in the list now to be given of false presentations, or are modified by some slight complication or peculiarity. They are, then, as follows:—

First.—One fore-leg only presented with the head lying upon it. In this case it is difficult for a ewe to lamb without help. The operator will endeavour to get hold of the missing limb, and, bringing it forward into its proper position, deliver the ewe. The best manner of doing this we shall consider after passing in review the principal abnormal presentations.

Second.—Both fore-legs lying back, the head alone being presented. In this position the ewe must have assistance, as birth without it is impossible. The head must be pushed back, the legs brought forward, and the lamb extracted.

Third.—The head slipped down between, or on one side of the fore-legs. This must be set right by bringing the head into its natural position above the fore-legs, and extracting the lamb.

Fourth.—A broadside presentation, in which case the broad side of the lamb is found within the *uterus*, and of course no progress can be made until the hand and forearm of the operator are introduced, and the *fœtus* is turned and brought into position.

Fifth.—The *fœtus* on its back, in which case a similar manipulation must be employed as in the last case.

Sixth.—A breech presentation. If the hocks are doubled, the breech of the lamb must be pushed forward and the hind-feet brought up. The lamb is then pulled away backwards without turning.

Seventh.—The *fœtus* too large, or the passage too small. This is a troublesome case, sometimes involving the loss of the lamb and occasionally of the ewe. Shepherds sometimes are obliged to carefully introduce a knife and cut off the shoulders and remove the *fœtus* piecemeal. More commonly

by patience and by exerting a good deal of strength the lamb
is safely born.

Eighth.—Monstrosities are not uncommon, most seasons
providing examples of lambs with five legs, headless lambs,
fusion of two lambs into one, &c. These cases are puzzling,
and require special treatment, and when such malformations
are presented there need be no hesitation in employing the
knife for their removal.

Having given all the possible unnatural presentations likely
to be met with, I shall next explain how assistance ought to
be rendered to a ewe in distress. In all cases great care and
gentleness are requisite, and all roughness or hurry should be
avoided. The hand should be anointed with fresh lard or oil,
and the finger nails must be short. The hand must be com-
pressed into as narrow a space as possible and gently introduced. In giving assistance the operator should draw the
lamb in accordance with the natural pains of the ewe, and
wait for her to pain. Assistance given at that moment is
useful; but if force is used during the intervals between the
labour pains, the muscles of the *uterus* are excited, and the
result is the early exhaustion of the mother. Again, in using
force the *fœtus* should be drawn downwards towards the hocks
of the ewe, and the operator need not be afraid of using his
strength when the *fœtus* is once brought into a proper position.

Casualties may always be expected in a large flock, but it is
only reasonable to expect that a shepherd understands his
business, and that he will be able to cope with the various
cases above cited.

After hard labour the parts should be soothed by suitable
applications. Decoction of poppy heads, turpentine, weak
solution of carbolic acid, and oils have all been used, no doubt
with good effect, as a means of preventing inflammation.

Too often, after ewes have had a severe time, and much
manual assistance has been necessary, "heaving," or after-
pains, accompanied with inflammation of the *uterus*, set in on
the second or third day and end in a painful death. When a

case of "heaving" occurs, the shepherd should not handle other ewes without careful disinfection, as, like milk-fever, the malady is propagated by contagion.

MORTALITY AMONG EWES.

The mortality among ewes after lambing is, it must be allowed, high, no one thinking that 5 per cent. of deaths is a matter for serious complaint. Three deaths to the hundred would, in most sheep districts, be considered as fairly good luck, and a few deaths in a flock are almost inevitable. These proportions are unquestionably high in comparison with the proportion of losses among dairy cows, but it is doubtful if they are to be reckoned as among our preventable losses. Sheep are exposed to such acute changes of temperature and of weather that the entire system of field management would need to be revised in order to prevent " chills " from producing their usual effects. In flocks which are proverbially "lucky" at lambing time—and there are such flocks in most localities— the immunity appears to be due to careful general management throughout the year, and to a dry sound soil, rather than to special skill on the part of the shepherd as to doctoring. Seasons evidently affect mortality in a manner which cannot be avoided, and in well-managed flocks, where the losses are heavy, it is generally found that others are suffering in a similar manner. On the other hand, there is, it would seem, an element of capriciousness, bad luck sometimes following good management, and bad luck suddenly giving place to a better state of things, as though the cloud of evil fortune had passed over. In sheep farming a bad beginning often is followed by a good ending, and just as we have begun to fear that things were going wrong altogether, the sun has seemed to shine out upon us again. Without doubt these views are unscientific, because there is always a *cause*, which it is our duty to investigate. But, if unscientific, they are, we believe, consistent with fact ; and until we know more of the treatment

of sheep in disease they are not likely to be falsified.   It is also
right to take into account that sheep must of necessity be
treated collectively rather than individually.   The best treatment for 500 ewes may be known, but the peculiarities of the
digestive system, or nervous system of each sheep cannot be
ascertained as it may be among cows and horses.   The human
doctor studies the idiosyncrasies of each patient ; but when a
veterinary surgeon is called in to a flock of sheep he is obliged
to prescribe generally, and the constitutional differences of
individuals must tend to render the results of his treatment
conflicting and unreliable.

SOUTHDOWN RAM.

CHAPTER XIV.

ORDINARY TREATMENT OF LAMBS.

THE management of lambs from birth to weaning is a subject of great importance. That it varies much, according to locality, must be freely admitted, from a method which may be summed up as a daily or twice-daily inspection, to the highly artificial system of close folding and constant changes of food, such as prevails upon the downs of Wiltshire and Hampshire. Between these extremes lie many variations in practice, differing in comparative expense and trouble; some very simple, some more complicated, so that it would indeed be difficult to please all parties by any description of lamb management.

Take, for example, the treatment of lambs in Northumberland or Lincolnshire, and we shall find, so far as food is concerned, that almost all that is required is abundance. Lambs are dropped in March and April, and after the usual care bestowed during the first three or four days after birth, lambs and dams are placed on fresh young seeds, or, it may be, upon permanent pasture. There may be seen Leicester or Leicester-Cheviot ewes with their lambs, the very picture of contentment, feeding upon the fresh herbage of Italian ryegrass and springing clovers, while their offspring play around, or make a rush upon their patient mothers two at a time, wagging their tails and "dunching" at her with their heads until they almost raise her hinder parts off the ground. In these cases the ewes are thinly run over the entire field, and changed from pasture to pasture in accordance with the supply of food. Corn may be given in small quantities when green food is scarce, or in the case of twins, or it may be as a little extra indulgence for two-teeth ewes or gimmers. The main idea is, however, that of grazing and growing without artificial help. A few thatched hurdles are put up, and a little straw is laid down around them for shelter, and the master or the shepherd visits them from time to time. Such visits I have often paid in years long passed away. The duty of the visitor is first to count the sheep, and this he does rapidly in twos and threes, as they dot the field singly, in pairs, in triplets, in little constellations, if the expression may be allowed, in which the groups seem to arrange themselves to the practised eye. The lambs are more difficult to count, as they are often hidden by the bulkier forms of their mothers as well as by hillocks, troughs, or other protections from the cold winds prevailing at this season. Lambs are fond of racing, and are not easy to count as they gallop the course they have chosen, and skip and play on any dirt-heap, or natural mound, or vantage ground. To notice if all are " full "—to attend to any little individual requiring a drop of cow-milk—a bottle of which, nicely warmed, the visitor carries with him ; to see if the tails are all free and not stuck down with hard dung— these are the principal objects of the inspector, who, having satisfied himself on these points, and seen that no ewe is lying awkward in a furrow, and that no lamb has twisted himself up in a sheep net, and in a word that all is right, walks or rides away. Twice a day or oftener should the flock be visited, and thus weeks pass by and the green foliage of April and May give place to the browner pastures of June and July, the sheep still finding their own living among the bents, and dry, but nutritious, grass. After weaning, the lambs still follow their dams, accepting their guidance, and thus they are carried on until turnips swell into substantial winter keep.

In the most recent account of Lincolnshire lamb management, given in the report upon the Nottingham prize farm competition, a similar picture is drawn of the summer management of breeding flocks, and if this were all, but little would be left to write upon with reference to feeding. Our attention would need to be directed to those episodes in a lamb or a sheep's life, such as docking, castrating, washing, clipping, weaning, and dipping. Also to such topics as maggots and foot lameness, which always need close attention. A pen should be made in a corner, and a dog is required to go round the sheep and bring them into close quarters, where the shepherd can catch and examine them. His shears are produced to cut off the wool where the maggots cluster thickly, and the mercury stone, or the Cuffs fly ointment, or other remedy, is applied to kill the wriggling pests. In other cases, he pares and dresses the feet, and applies the particular astringent which he favours, be it a solution of sulphuric acid, butter of antimony, or a paste bought of the druggist. Flies attack both head and tail as well as any other part of the carcase, especially when it is damp or dirty. Nothing can be more painful and irksome to a sheep of the Leicester-Cheviot breed than the pertinacious settling of flies upon his devoted head. They form a black helmet between his ears. They suck the juices from his scalp, trample it with their feet, and by the friction of their bodies and wings soon wear a bare place into a scald upon which they revel. The unfortunate animal in vain tries to shake them off. He runs and stops, he stamps with his feet, waggles his ears, shakes his head, squats among the bents, or seeks the shade, and lies as close as a hare in its form—but all in vain. In this sad plight he is found by the shepherd, who, summoning his faithful quaint-looking dog to his aid, drives him and all his fellows to the pen, and proceeds to make him comfortable. If the injury is superficial, a dressing of sulphur and oil applied with a brush is sufficient; if the skin is raw and broken, the sheep cap upon a pad of lint, supplemented with a dressing of Canada balsam upon the sore, may be applied.

A shepherd should always carry a knife with a blade sufficiently long to slaughter a sheep if necessary. Such meat is not generally thought to be unwholesome by rural folk, and if the carcase dresses out fairly well, it finds a ready sale among the work people; and the farmer himself is not averse to taking a quarter for his own household. In the light of modern science this appears wrong, as blood poisoning, car-buncle, and other evil consequences have been traced to the consumption of diseased meat. On the other hand, thorough cooking appears to be a fairly good safeguard against such evils, and all we can advise is that when braxy mutton is eaten it should be well roasted or boiled. It would go sadly against the instincts of a true shepherd to bury a carcase slaughtered to "save its life," or rather its value; and the master also would think himself greatly wanting in thrift if he followed high medical advice, and sacrificed a sweet bit of meat obtained under such circumstances.

Such is an outline of the management of lambs, as carried out in many counties. It contrasts forcibly with the next picture which I shall endeavour to present, of constant change of food, corning from birth to death, and the production of lamb prodigies. The two cases are so different that descrip-tions of the latter system has often aroused criticism from sheep-breeders who follow the less sophisticated method just described. According to the first plan lambs in August will be worth from 15s. to 30s. a head; and as an extreme case applying to black-faced lambs 6s. may be mentioned as a satisfactory price.

I shall next endeavour to describe a system adapted for producing lambs of 20lbs. per quarter at eight months old, or rams ready for service at the same age. It is well, however, to point out that the difference between the management of a lot of lambs upon a Yorkshire fell or Northumberland hillside farm is very different from the system pursued upon a Hamp-shire farm with its double cropping of fodder crops and roots, and its diversity of foods for each month of the summer. On these farms the lambs frequently are indulged with three or four separate kinds of green food every day, and they often spend no more than three or four hours upon one particular fold. Rape, clover heads, cabbages, and vetches, with cut mangels, cake, and corn constitute their daily fare, and thus growth is extraordinarily rapid.   It is no exaggeration to say that during the height of the season lambs will increase in live weight at the rate of one pound a day—an addition which must appear marvellous, or past belief, to many graziers of sheep.   The full details of the system must furnish material for the next chapter.

LINCOLN RAM HOGGS.

CHAPTER XV.

EXTRAORDINARY TREATMENT OF LAMBS.

I TRUST that those of my readers who have followed the last chapters upon Sheep and their Management will not judge me harshly if I now proceed to take as my special theme my own breed of sheep.

It is as lambs that they particularly excel, and I have often declared a Hampshire Down ram lamb, as he appears in the sale-ring at the Market House, or on the Butts, at Salisbury, late in July or in early August, to be one of the wonders of the world. These lambs are for the most part born between the 10th and the 31st of January, and the principal fall is about the 20th of that month. A few are dropped late in December, but this is not thought desirable by ram-breeders, as they are then apt to have passed their most perfect bloom before they are disposed of. Nothing, we know, beats a January lamb ; and, if we take January 20th as the day upon which a lamb is yeaned, we shall be able to show a record of a daily increase in live weight of $\frac{3}{4}$ lb. from the day of birth to August 1st—that is, a lamb of 144 lbs. weight at 192 days old.

Comparing this with the increase of ordinary sheep or even of cattle it is very striking, and will scarcely be credited by those who are accustomed to the usual system of bringing up lambs, described in the last chapter. The result shows the wonderful earliness of maturity in improved Hampshire Down sheep as a breed, in which quality they are unrivalled, and also the merits of the system of feeding which can produce it.   The three factors necessary for the achievement of such a result are, first the breed, second the mode of feeding, and third the peculiar soil and climate of a southern county adapted for the growth of summer fodder crops as well as of good root crops.

To bring out a 12 stone lamb at eight months old is within possibility in the case of this breed, and yet this was the limit I proposed when contrasting the usual management of sheep with what I am now about to describe.

A Hampshire Down lamb may then be supposed to be born on January 20th, and we have to follow his short history until he appears either as a wether ready for the butcher early in August, or as a ram lamb fit for service at the same date.

In the first place it will be necessary to feed the dam liberally in order that her milk may be both plentiful and rich, and with this end in view we early begin to feed with cake, giving an allowance of 1 lb. per head per day.   This, together with hay and turnips or swedes, constitutes her diet, and this is continued for at least ten weeks, or until such time as it is considered advisable to lower the amount of cake or corn given to the ewes in order to increase that given to the lambs.   Both ewes and lambs are comfortably housed at night in a well littered and well sheltered pen, and have daily access to a fold of turnips, and receive their cake and hay regularly.   The young lambs quickly learn to nibble at the turnip-tops and to select the finest portions of the hay, and when this is noticed it is time to give them a corner to themselves, where they can have a little finely-ground linseed cake, split peas, oats, and crushed malt.   This they soon learn to relish, and it is pleasant to see them crowding around their troughs after their corn, and then passing through the creeps or lamb-hurdles to steal a drop of milk from their mothers.

The lamb hurdle is from this time an institution.   By its

means the young creatures can run forward and crop the first green food of the season in the form of succulent swede or rape tops. They are at this time in receipt of eight different sorts of food—namely, hay, turnips, turnip or rape greens, linseed cake, peas, oats, malt, and milk. Their progress is wonderful, and their short and smooth coats evidence their perfect health. As the season advances into March further change is obtained by an outrun on young grass for three or four hours in the middle of the day. This system is pursued until early in April, when they enter upon a succession of spring and summer fodder crops which were sown extensively the previous autumn. This series begins with rye, over which swedes or mangels have been heaped, and from this fold they go daily in many cases into water-meadow, always returning to the rye-fold in the afternoon. These sheep are always between hurdles throughout the year, although a "spread" over a field is frequently allowed during a few hours of the day. The lambs also continue to enjoy the privilege of choice of food by means of the lamb hurdles, so that while the ewes are close-folded the lambs enjoy much greater freedom. There are generally two descriptions of green food going at once, and, in many cases, three or four, according to the season of the year. Following the order in which the fodder crops mature, we find that these lambs are, successively, fed upon turnips and rape, rye and water-meadow, winter barley and water-meadow, winter barley and trifolium, trifolium and vetches, vetches and rape, rape and cabbage. There are at least two distinct changes of food every day, and as vegetation becomes more luxuriant they often have a variety of courses which is quite epicurean in its character. Take for example a fine midsummer day, when the lambs awaken upon a fold of vetches. The shepherd is up betimes, and begins by giving them an allowance of cake. He then grinds some mangel into troughs, which they eat with great relish. They are next admitted to a fresh fold of vetches, after which they are walked quietly away to a neighbouring piece of good rape or cabbage. After two hours or more, and in the heat of the afternoon, they are allowed to spread themselves over some old sainfoin or aftermath clover. They will then return to the vetch fold, and after receiving another feed of corn they lie down to well earned repose, having increased their weight by 1 lb. each. Hay chaff in troughs is also frequently supplied, even in summer, by way of keeping them firm in their bowels; thus, a lamb may easily partake of eight different kinds of food. Rape and cabbage or kale give way to turnips in late July or early August, and the allowance of "corn" is kept up to from 1 lb. to 1½ lb. per head. This allowance is pretty constant from birth, considering the cake given to the ewes, which is, of course, supplied for the benefit of the lambs. Weaning usually takes place in May, and the ewes then go on to hard keep.

Such is the system, so far as it can be described, by which these lamb prodigies are produced. The most serious expense is probably the cake and corn, but the total amount consumed per head is not so serious as might at first be thought. At the very fair allowance of 1 lb. per head per day for the entire period of the lamb's short life it would not be more than 200 lbs., and 2 cwt. would probably give all that is required. At 8s. per cwt. this equals 16s., so that we are probably justified in stating that the corn and cake costs well within 20s. a head. The increase in value due to the cake is, I submit, considerably beyond these sums, so that it appears probable that these lambs pay for the cake. Lambs treated according to the system just described have been sold at 64s. each on August 1st, and could not have been worth 44s. apiece without cake and corn, so that we have good reason for thinking that the cake and corn is paid for, and more than paid for, by the increased value of the sheep.

Careful shepherding, suitable land, plenty of change, liberal allowance of concentrated foods, and a good breed to work upon, are the chief points required in order to secure success. The remaining points of management are the tailing of the lambs, which is done by searing at a month old; castrating, which is usually performed in May; and careful attention to the feet in order to prevent foot lameness.

DOCKING.

Lambs should be tailed when about a month old, and it is better to castrate at a later date. Docking or tailing in necessary for several reasons. A sheep's tail is sadly in his way. It is a feeble member, quite incapable of being used for brushing away flies or for any other object. It is a useless appendage, although it, no doubt, at one period of sheep history served as useful a purpose as the tail of other animals to them. But the development of wool has altered the sheep from his original form, and now the tail is too liable to become filthy from dung and mud, and it is better off. The operation is simple and instantaneous, and lambs begin sucking immediately after the severance has been effected. It may be done in a moment with a knife, but less blood is lost, and healing is promoted, by the use of a hot sharp iron which is pressed through the tail at the required point as to length. This may be close to the rump, or about three inches lower down, according to fashion. The tails may be made into pies, and are esteemed by some as a delicacy. They should be steeped in boiling water, and the wool then readily comes off, leaving the tails bare and clean. Lambs look all the better for tailing, and a collateral advantage is that when fat they handle better at the dock or rump, than if they carried a long tail. Docking should not be attempted in frosty weather, as the frost seizes on the wounded portion and prevents it from healing.

CASTRATING.

The knife may be employed for castrating. It is very common practice to castrate and tail at the same time, but the treatment is severe. I prefer to postpone castration until the weather is mild and the ground warm, as it is dangerous for a newly cut lamb to lie on the cold earth. The objection to this course would probably be raised that in warm weather flies abound, and attack the scrotum and possibly breed maggots within it which would of course be a serious complication. The best mode of avoiding this is to use the actual cautery in the form of an iron heated to the proper temperature. The scrotum is cut open with a knife and the testicles are brought forward until the cords can be grasped by a pair of clamps. The testicles are then seared off close to the clamp, and, before the cords are released, they are thoroughly cauterized and dressed with lard and a proper application, and returned into the scrotum. There is no bleeding, and the clamp probably deprives the parts to be operated on of sensation. It is seldom that any loss follows the castration of ram lambs if the operation is skilfully performed. It is usually done by an itinerant expert, who travels from farm to farm, earning capital pay during the two or three months in which this work is generally performed.

Sheep Washing.

The yolk, or yelk, is an important item in wool. When we remember that in the process of washing the fleece loses from one-third to one-half its weight through the loss of the yolk, it is evident that the question of washing or not washing is of vast importance. It will be in the memory of readers that this question has been raised of late, and that farmers have been recommended to shear their sheep in the grease, and accept the lower price which the wool commands under these circumstances. Mr. Turner, Bradford, appears to think that wool is best washed in cold water. He writes as follows: " I saw an article on ' Wool ' the other day in which the use of hot water and soap was advocated. With this idea I entirely disagree. I venture to say that the use of hot water or soap is positively injurious to English wool. What I would recommend (of course with all reservations as to

convenience) is tub-washing in cold water." Between such
a method and the ordinary system of pool washing there
is no wide divergence, and hence there does not appear to
be any urgent necessity for changing a system which has
been found to work fairly well. At the same time, if tub-
washing could be substituted for pool-washing, there is the
advantage of the presence of the yolk, which is a true and
natural soap, and therefore well calculated to assist in the
thorough cleansing of the wool. The following manner has,
in fact, been found satisfactory, and might be adopted when
circumstances are favourable to its use: " Having two tanks,
each capable of holding, say, five sheep, let the sheep be
placed in tank number one and washed in the usual manner ;
then let them be plunged in the second tank, which must
be kept constantly supplied with clear water. It should
be remembered that the first tank should be kept as greasy
as possible, only as much water being added as the sheep
take out with them. The yolk is a kind of natural soap,
and is quite sufficient to wash the sheep properly if advantage is taken of it." This system has the advantage
of great simplicity, and is really only a modification of the
usual practice. Given a good water supply, all that is
needed is to deflect a stream so as to give a sufficient supply
of water to both tanks or sets of tanks. For example, the
stream would supply a regulated amount of water for keeping
up the level of the washing water, while a continuous flood
could be easily maintained through the lower or second series
of tanks. The sheep, after being washed thoroughly in the
upper tank, would be handed on for a plunge or swim in the
lower tanks, and be allowed to land and shake themselves.
Such a system will commend itself to practical men as
feasible, and might be at once adopted without much outlay.
There is indeed no reason why one set of tanks only should
be constructed for washing, and that the sheep should then
be thrown from the greasy water into the burn or brook, and
allowed to swim out.

The system of going into the water for the purpose of washing sheep is more prevalent in the north of England than in the south, in spite of the greater severity of the climate. Sheep washing in mid stream is no doubt a trying ordeal for a weak constitution, and is a cause of rheumatism. The upper part of the body becomes over-heated, while the lower portion is immersed in the running water, and in no circumstances is a drop of whisky more advisable or excusable. We do not know of any harm having happened to sheep washers during, or just after, the work, but it is probable that it may sow the seeds of future ailments. The advantages of the system is that the sheep are washed while on their backs, and that the dirt and sand are more easily washed out and fall downwards by gravity. Swimming sheep through a pool is the alternative, and in this case the men are furnished with T-headed rods resembling hay rakes without teeth, and with these the sheep are scrubbed and soused under the water until the operators are satisfied, and the sheep is then allowed to land. It is possible that the prevalence of short-woolled Down sheep in the southern counties, and the greater distribution of long-woolled sheep in the northern counties, may be the reasons for this slight difference in practice.

Sheep ought to be washed ten days or a fortnight before clipping, so as to allow the yolk to once more rise in the fleece, and thus add to its weight, its suppleness, and lustre. Hot weather is most favourable for this important object, and washed sheep should be placed on grass land free from sandy banks, under which the animals might rub and thus collect sand on their backs.

In view of the fact that a large proportion of foreign wool reaches the London market in the grease, it is only reasonable that farmers who are favourably situated should put to the test the advantages and disadvantages of washing their sheep.

SHEARING.

Towards the end of May shearing will commence. Some time ago I witnessed an interesting display of shearing with the Wolseley machine shears. Having heard of the favour with which this instrument is being received in Australia and New Zealand, I took the opportunity of a public exhibition of this method to become acquainted with the procedure. The process was most successful and rapid, and may be described in few words. The power used was a 1½-h.p. vertical steam engine, from which a belt actuated the shears. The power was conveyed through a flexible tube, similar in principle to what is used by dentists for cutting into decayed teeth. Some of us have experienced the operation of tooth-stopping when the swiftly revolving file or rotating cutter is applied by a skilful dentist. Similarly, the shearer holds in his right hand a machine which resembles in its structure both a horse-clipping machine and the knife and guards of a reaping machine. It is a compact, workmanlike-looking instrument, and when actuated by the power the knife oscillates between the guards with immense velocity. The sheep is held and handled as in ordinary shearing, and seems to yield itself with passive indifference to the mechanical novelty. The head, cheeks and throat are freed from loose wool, and the fleece around the ears is removed. The breast and belly wool is then opened out, and in a few broad cuts the fleece begins to peel off rapidly under the hands of the shearer. The work is superior to that usually done by hand, and much more rapid, as a good operator turns off a sheep every seven to eight minutes. The wool is more cleanly sheared off, and there are no ridges left of longer wool. The danger of cutting the skin seems also to be slight, and in the specimens which I saw shorn there was not a scratch or cut to be seen. The cost of the implement, with its tubing, bracket, and pulley for attachment to the power, is £10, and a man is sent out with each instrument to instruct beginners in its use. I was informed that any man could learn how to use the shears in half a day.

How far this implement is likely to supersede our slower and more laborious method it is difficult to say, but perhaps the comparatively small numbers of our sheep will militate against its extended use. The smaller size of the colonial sheep has also been in favour of the Wolseley machine shears, and the promoters of the demonstration already noticed evidently were a little disappointed at the area of skin to be got over in the case of our large sheep. I see that the average tally of one man with these machines is stated to be 115 per day, while a maximum of 203 is said to have been reached. At the rate at which the work progressed with good English tegs, I should say that a very good average result would be eighty sheep in a day of ten hours.

The Newall-Cunningham Syndicate, 73, Cheapside, London, and Burgon and Ball, La Plata Works, Malin Bridge, Sheffield, also supply machines for the same purpose at £10 to £15 each, which satisfactorily shear sheep.

Weaning.

Many years ago an excellent paper on sheep management was contributed to the Journal of the "Royal" by Mr. Pawlett, in which he advocated weaning lambs at eight weeks old. I have often thought about it, and how Mr. Pawlett found that even in the succeeding February his early-weaned lambs were heavier than those which had sucked later. A good deal must depend upon the season at which a lamb is born, for to those who lamb down their flocks in January, weaning in March appears absolutely impossible. Those sheep farmers whose flocks lamb down late in March are differently situated, and the advent of summer weather, and the presence of plenty of green food, may render weaning possible early in June, or when the lambs are from ten to twelve weeks old. So much is done by rule on most farms

that perhaps few flock-masters trouble as to the precise age
of a lamb when it is weaned.　One may be a full month older
than another when the order is given, and yet both must
share and share alike.　" The season brings the flower again,
and brings the firstling of the flock," and on as near the
same day as possible the weaning takes place, as also does
the washing of the ewes and their shearing.

Gradual weaning appears the best for both lambs and dams.
The severe system of taking away the ewes and placing them
on dry keep out of hearing of their bleating lambs is a plan
which is followed, especially when open grazing is the rule.
When sheep live much within hurdles and the lambs are
allowed to run forward before the ewes, a simple means of
weaning is to stop the runs for a few hours each day, and thus
gradually accustom both lambs and ewes to live without each
other.　The lambs in this case have the first choice of the
" victuals," and the ewes may be held back on somewhat hard
keep, " eating the crusts," so as to dry up the flow of milk.
When treated judiciously little inconvenience seems to be
suffered on either side, the lambs being allowed to draw the
udders at increasingly long intervals, until they are weaned.

CHAPTER XVI.

SINGLE AND TWIN LAMBS.

THE number of twins or of single lambs is an important matter affecting the profits of sheep farming. An abundance of twins is a matter for congratulation, but is not an unmixed advantage. They will not attain the size of single lambs for sale in the following autumn; the ewes require more food, and are often more reduced in condition through suckling, and the strain upon the mother is heavy, especially in the case of twotooths. Still, a good many twins are required in order to keep up the number of lambs, which is liable to drawbacks from death, barrenness, and slipping. Twins give the opportunity to the shepherd of dividing them, and thus supplying lambs to ewes which have lost their own offspring, and which, otherwise, would go as barreners. Without a fair proportion of twins we should unquestionably suffer from a short supply of lambs, even upon the assumption of a lamb to a ewe throughout the flock. This apparently modest estimate is by no means always realised, in spite of twins, as barren and aborted ewes may easily constitute 5 per cent. of a flock, and often double that proportion. Deaths among very young lambs are also frequent, so that the general statement that " for every ewe put to the ram there should be a lamb at weaning time " is not far from correct.

How to Obtain a Big Crop of Lambs.

Some flocks, and some farms, seem naturally adapted for producing a large number of lambs. It may be reasonably

expected that twins will in turn produce twins, and hence
rams and ewes which have been twins might properly be
selected to propagate their species.   Fertility is as likely to be
inherited as any other property, and with it the natural accom-
paniments of good nursing and abundant milk supply.   I am
inclined to think that ewes are naturally disposed to produce
a pair of lambs, and that single lambs are to be regarded as a
degree less normal than twins.   Thus, when ewes are in good
order and keep is abundant—both of which conditions must
be regarded as strictly natural—the number of twins is immediately increased, and sometimes almost the whole flock produces doubly.   This indicates the best method of obtaining a
big crop of lambs, namely, keeping the ewes well throughout
summer.   Extreme fatness or extreme poverty both militate
against fertility, but a judicious mean and plenty of good food
during the period of conception produce an opposite effect.
Ewes which have been barren during one season will often
conceive early and produce two strong lambs the succeeding
spring, and purchased ewes which have been caked will
generally produce a lot of lambs.

Young Lambs.

The first duty of the shepherd after a lamb is born is to
clear its mouth of mucus, and see it draw its first breath.
Previous to birth the *fœtus* receives oxygen through the
mother.   It is her lungs which vivify its blood, and her
digestive system which prepares its nourishment.   But with
the breaking of the *umbilical* cord comes the necessity for air,
and after a convulsive movement of the diaphragm and inter-
costal muscles the young creature gasps, and generally utters
its first cry.   Whether the almost universal practice of shepherds of blowing into the lamb's mouth facilitates this action
is not certain, but it is probable that this simple expedient
excites the slumbering vitality, and causes the necessary
muscular contraction.   A slap with the flat of the hand across

the buttocks will also often cause a lamb to draw its first breath, when animation appears to be suspended for a few seconds after birth.

The lamb is most conveniently carried by its two fore-legs, and in this position it is taken to one of the pens, or cribs, already described, followed by its anxious mother. As soon as the lamb has got on to its feet, and has found the teat, and the shepherd is satisfied that the ewe has a sufficient supply of milk, he will proceed to give her two or three swedes, a mangel, or better still, a white or yellow turnip or two. A little hay may also be supplied, and after these simple attentions the ewe may be left in charge of her lamb, and if all goes on well there is no need for more elaborate treatment.

From what has been said it is then evident that where a ewe is naturally delivered of her lamb she requires no medicinal treatment, but only to be placed in a sheltered position and to be fed with ordinary food.

DIFFICULTIES.

One of the first difficulties to be overcome happens most frequently in the case of two-tooth ewes, which occasionally refuse to take to their lambs. This is more likely to occur in the open field than in a pen, and the best plan is to place lamb and dam in a small crib and hold the ewe while the lamb sucks. In all such cases a little patience is all that is required, and in a short period the maternal instincts will assert themselves.

It is less easy to make a ewe take to a strange lamb, but this is managed by the shepherd without much difficulty. One of the best methods is to keep the lamb in readiness, and as soon as a ewe with plenty of milk is delivered of a single lamb the stranger is rubbed with the newly-yeaned lamb, and smeared with the liquor *amnios* which flows copiously from the womb of the dam. The lamb is then presented to the ewe, which will at once take to it, and place it on the same affectionate footing as her own natural offspring.

If a ewe loses her lamb, and it is requisite to replace it, one of a pair of twins is generally selected, or it may be a lamb which has lost its dam. The skin of the dead lamb is removed, stretched over the little stranger, and made secure. The ewe is then put in a small coop or crib and placed in the "stocks," and the strange lamb will then make itself at home, and in a day or two the ewe's objections will disappear, after which she may be released. It may seem unnecessary to further describe the process, but as some of our readers may be ignorant of what is meant by the stocks, I will briefly describe what is meant. Two hurdle stakes are driven firmly into the ground, about 7in. apart, and a thong or shackle is passed over the tops of them both. The head of the ewe is then put through the stakes, so as to hold her by the neck, and the thong is used to keep the stakes close enough together to imprison the ewe. A third stake is passed horizontally under the ewe's belly, and supported at the two ends on the bottom bars of two hurdles placed on either side of the ewe, but at a sufficient distance to allow the lamb to approach its foster mother. The ewe is thus not only held but is unable to throw herself down to prevent the lamb from sucking. After a short discipline of this kind the ewe will generally take to her new charge.

When ewes have not a sufficient supply of milk, cow's milk may be used. Various views have been expressed as to whether a cow recently calved or old in milk is most suitable for supplying nourishment to a young lamb. In Wilson's "British Farming" we are told that to give a lamb milk from a newly-calved cow is tantamount to knocking it on the head, but the experience of many shepherds is entirely contrary to this forcibly expressed opinion.

When a young lamb is weak and unable to stand, it should be placed near a fire, and a teaspoonful of gin in a little warm water, sweetened with moist sugar, may be administered with good effect. The ewe should be milked into a cup, and the lamb fed with a spoon, and in a few hours it will probably regain its strength, and be able to rejoin its dam.

When lambs are affected with diarrhœa a teaspoonful of castor oil will remove the acidity which is the cause of the disorder, and check the complaint.

When lambs are two or three days old they are liable to a rapidly fatal form of diarrhœa, known as the "scant." When once this disease appears in a lambing pen it is difficult to cope with it, and the loss may become serious. Instead of attempting direct treatment, the best system is to make a new pen at a distance from the old one, and to remove the ewes and lambs to it. By this change of lair and of food the disease is usually stopped. As the scant only attacks extremely young lambs, this practice entails less trouble than might at first sight appear ; it is, in fact, only necessary in the case of the newly-lambed portion of the flock and those expected to lamb. The pen should at all times be kept healthy by the use of plenty of litter, and the removal or burying of all dead lambs or ewes, and all cleansings. This is too much disregarded by shepherds, but it is one of those points which a master may properly enforce.

EARLY TREATMENT OF LAMBS.

The treatment of very young lambs differs according to the earliness or lateness of the yeaning time. When flocks lamb down in March and April, and where the close folding system gives way to the more natural plan of open grazing, the ewes and lambs are first turned out on to pastures or seeds. In the close folding districts of the chalk formation, where lambing principally takes place during January and February, greater care is requisite, and a more artificial system of management is pursued. In this latter case ewes and lambs are turned out upon turnips or swedes. During the first fortnight lambs subsist entirely upon milk, but even before the lapse of that short period they will be seen playing with the finest portions of the hay and the turnip greens. When it is desired to bring out a ot of ram lambs or early wethers at eight months old, no time should be lost, and after the first fortnight an enclosure of hurdles is made for the lambs and furnished with suitable troughs.

LAMB HURDLES.

At this stage the lamb-hurdle, or lamb creep, becomes an important institution. It is a contrivance by which the lambs are allowed to run forward while the ewes are kept back. The same object may be attained by disposing ordinary hurdles so that the lambs can pass out of the fold at pleasure. The lamb-hurdle is in constant requisition throughout the spring, and by its means the lambs are able to run forward and crop the choicest herbage before it is soiled or trampled by the older sheep. The best creeps are adjustable to the size of the lambs, and the upright bars through which the young animals pass are round and smooth, and revolve easily upon a central axis of iron. They are also furnished with a similar roller, which forms the top of the creep, so that the lamb passes through without rubbing the wool. The opening is hinged inwards, but is rigid when pushed outwards, and this is done to allow of lambs running quickly back into the fold if frightened, but at the same time to prevent the ewes from passing outside the fold.

SHEPHERDS.

Those who employ shepherds must find it entertaining to talk to them. Let a master be ever so well practised in the arts of farming, or ever so well read in the subject of sheep, he will find his shepherd's remarks (that is, if the shepherd is worthy of his class) full of freshness, and of immediate practical value. I have studied shepherds in many sheep districts, in England and in Scotland, and believe them to be a genuine, sound, and reliable set of men. The hill shepherds of the Highlands are men of great experience and trustworthiness, as are also those of Lammermuir and Cheviot. In every sheep

country they form a distinct class, and, on enquiry, you invariably find that they are the sons of shepherds. Probably if the inquiry was pursued, it would be found that the occupation is for the most part hereditary, and that shepherds form a sort of caste among themselves. It would indeed be difficult to make a shepherd out of a man who had not from his earliest years and associations been accustomed to follow the flock, and hear all that was said about them by his father and brothers. The art is imbibed with the mother's milk, and as the boy grows he learns to know all about sheep instinctively, and as a second nature. A true-bred shepherd probably requires less training than would a town-bred lad, however early he might be apprenticed to the trade. The knowledge is not always easy to fathom or understand. If a shepherd says a ewe will die, she is only too likely to die. If he says she will get well, he is almost invariably right. He knows when she is going to have twins, and he shrewdly suspects when she is about to produce triplets. His prognostics as to good and ill luck almost partake of the oracular, and his suspension of an opinion is ominous. He does the right thing as a matter of instinct, and knows what will agree or disagree with his flock as if he were a sheep himself. When the master suggests a change of food, the shepherd, as in duty bound, assents; but, as also in duty bound, expresses his opinion that the sheep will "scour" or "blow," or go back, or do something which is not altogether desirable. Experienced masters in such cases allow themselves to be guided a little, and modify their orders. Inexperienced masters sometimes learn to their cost that the shepherd was right. If the master knows better than the shepherd about every detail, then the probability is that the shepherd is not a very good one. It can, indeed, hardly be expected that the farmer, who is the administrator of the entire business of a large farm, should know the sheep as well as a man who is with them all day, and in certain seasons all night, and who knows each individual of the flock as well as a schoolmaster knows every boy in his school.

The memory of a shepherd for individual sheep is very remarkable, and it is striking to note how quickly he picks up this knowledge when he enters upon the duties of a new place. It is also curious to observe how rapidly shepherd boys under their father obtain this power of recognition and a memory of faces. The master admires a lamb, and the shepherd proceeds to speak of its mother, and the circumstances under which she was born, and to tell him that she was a late lamb, and at one time gave little promise of her future excellence. This other lamb, now growing into a fine specimen, and likely to be of value as a sire, was that same little helpless thing which the master had noticed, and told the shepherd that it would never come to anything; and this capital pair of lambs are from that ewe which broke her leg in falling over a chalk pit two years since. Or it may be that a ram lamb is chosen out of a lot of a hundred or more for a certain purpose. The master suggests marking it, but the shepherd protests that he will know him again, and know him he will, without fail.

In the same manner the shepherd's suggestions as to the description of corn or cake required to push forward his sheep for sale are sound, and worthy of attention; and he will not be led aside from his opinion by any quotable authority whatsoever. It must not, however, be inferred that shepherds are infallible or that masters are to follow their shepherds' behests entirely. A shepherd would make everything on the farm bend to the requirements of the flock, but it is the master's duty to see that all departments are equally well served. Shepherds also care nothing about expense. They want cake, they want hay and corn, the best pastures, the run of the best root crops. They do not spare the master in their request for hurdles, for troughing, for keep; but the master must maintain a judicious firmness throughout, while he hears all that his shepherd has to say on these subjects.

Shepherds are somewhat antiquated in their ideas of surgery, and are too liable to have recourse to bleeding and the use of the knife. In their treatment of foot rot, of bad udders, and of sturdy, they are guided by crude notions, and probably often do more harm than good by their treatment. While the fact remains true that the veterinary profession is, comparatively speaking, ignorant of the treatment of sheep, shepherds are likely to continue to discharge the duties of doctor as well as nurse. Drugging sheep is not usually successful, as they are much more likely to die under treatment than either cows or horses. You may bring a cow through an illness which at one time appeared in every way likely to prove fatal, but when once a sheep is down there is but little hope. So far as results go, the rough, and sometimes what seems to be the cruel, treatment of the shepherd is just as likely to be successful as that of the " vet." It is, indeed, not easy to get shepherds to follow the long list of directions given by a professional man, and faith in the benefit of what is suggested by him is too often entirely absent.

CHAPTER XVII.

WINTER FEEDING OF SHEEP.

A GREAT change has come over the winter feeding of sheep during the last thirty years. Thirty years ago I was introduced to farming, and I well remember that tegs were fatted in my part of the country so as to come out of their wool fit for the butcher at from fourteen to eighteen months old. They were usually shorn and sent to market as they ripened, being drawn one or two score at a time as they became fit. The same system no doubt prevails still, but nevertheless, there has been a marked movement in favour of early maturity. Some twenty years ago Mr. W. J. Edmonds, of Southrop, in Gloucestershire, read a paper, in which he spoke of bringing out eight stone sheep at ten months old. We now speak of bringing out ten stone sheep at eight months old. I can remember when Mr. Hulbert's nine to ten months old sheep made a sensation in Cirencester market when penned in November. Now seven to eight months old sheep, weighing 20 lb. per quarter, are constantly shown.

As further proof of the advance in earliness of maturity, it may be mentioned that even in the memory of middle-aged shepherds it was very common to keep back two-tooth or shearling ewes till they were four-tooth, or, as it was called, " double two-tooths," before they entered the ewe flock. Now many people advocate tegs dropping lambs at one year old.

A large number of lambs in these days never see winter keep after they are weaned. They are begun in January upon turnips, and are sold to the butcher in August, September and

October, at an age when at one time they would have been called lambs.

The winter feeding of tegs is now confined to certain breeds of sheep, such as the Cheviot, Leicester, Cotswold, and some of the Down breeds. It is also used in the case of the smaller and later lambs of our most forward breeds of sheep, such as the Hampshire and Oxford Downs, the foremost and earliest lambs being sold to butchers in the early autumn. The winter feeding of tegs, therefore, is most generally carried out in the case of breeds not famous for early maturity, and of later and smaller lambs which are bought at the autumn fairs.

Change from Summer to Winter Grazing.

Grass loses its " nature " in September, and during October tegs are finding their way from open grazing to the turnip fold. The change should be made gradually, by placing the tegs for a short period daily upon turnips and allowing them to back run upon seeds or stubble. By this system they become accustomed to winter fare, and by the time they are closely penned upon it they have ceased to pine after green food. The change from summer to winter keep should be made with care, as, indeed, should all changes of diet.

A few white turnips scattered over the pastures, or folding on the turnips for a few hours only at first, with frequent changes, tends to keep sheep healthy, and free from casualties. To pen at once upon swedes, hay and corn, would be highly dangerous, and certain to result in the death of some of the flock. No root is more suitable for early winter feeding than the Pomeranian and common white globes. They are palatable, easy of digestion, and not too forcing or heating. Up to December, sheep will do better on these soft turnips than upon the harder and richer yellow and Swedish varieties. Little hay is needed as long as grass can be had on the runs, and the stubbles afford grazing, but as winter closes in and the dark and wet days of November approach, something comfortable is needed, and for this purpose nothing is more suitable than hay. Clover hay is too rich in albuminoids to commence with, and besides, it is policy to cut into the inferior ricks first, and to reserve the best hay for finishing with in spring. Meadow hay and white turnips will be gradually changed for clover hay and swedes after Christmas. The lower temperature which will then prevail, and the more matured constitution of the sheep will at that time be favourable for the introduction of food of a more nutritious character. The differences in composition between these classes of roots and hay are very great, and a diet of white turnips and meadow hay is very different from one of swedes and clover hay. Swedes and clover hay are highly nutritious and heating, while meadow hay and white turnips, being poorer in albuminoids, are less heating to the system and easier of digestion.

DRY FOOD.

Too much stress can scarcely be laid upon the paramount importance of plenty of dry food for sheep.

Turnips are too watery for winter feeding when used alone. Grass is also watery, but the circumstances under which it is consumed are so different that its high percentage of moisture is not injurious.

It has been the habit of writers to decry turnips as consisting largely of water. It must, however, be remembered that all succulent and luscious vegetable growth is watery, and that even the animal body consists principally of water. Nature's most perfect food for stock contains over 80 per cent. of water, and yet no combination of artificial foods has yet been contrived to rival young grass. It is, however, the misfortune of the turnip to contain an excess of moisture and to be fed at a time of year when the very mention of cold water seems to send a shudder down the spine.

To awake on a piercingly cold morning on a bleak hill-side, and to partake of a breakfast of " ices "—for such would be

about the correct description of turnips in such circumstances
—does not seem inviting, and, in fact, it is extraordinary how
the animal system can stand such treatment, and in many
cases thrive upon it. The advisability of adding some food of
a more comforting and concentrated character is apparent,
and this leads me to speak of the various descriptions of dry
foods for sheep.

Winter may be considered to include December, January,
and February. Agriculturally, the winter is somewhat longer,
and yet November is scarcely a winter month to the stock-
feeder, and March brings a certain relief in the form of early
rye, young seeds, and in some favoured districts in the form
of water-meadow grass. July-sown rape, and late-sown turnips
also begin to throw up green fodder in March, so that some
time during that wild month we begin to feel the influences
of returning spring. Still, the stock-feeder's winter is considerably longer than that of the calendar, and we shall probably not be far wrong if we assume that our flocks rely chiefly
upon roots, hay and purchased foods, from November 15th to
April 15th, or for a period of twenty-one and a-half weeks.
As late springs are among the evils which we may expect in
these latitudes, it is necessary to provide a supply of winter
keep for a somewhat later period, to be used concurrently
with spring crops; and hence a sufficient stock of swedes,
mangel and hay, to last into May, must be considered as a
point in management.

This can be accomplished by two methods, the first of
which is the more certain and practicable. It is summed up
in a piece of sound advice to all sheep keepers, namely, not
to overstock. Overstocking is liable to produce an uncomfortable and perplexing dilemma. The winter's provision of
food becomes prematurely exhausted, and the pinch comes
just at the time when it is undesirable to sell. The case
stands thus. We encounter a late spring, and store stock is
found to be a drug. The consequence is that we must hold
until the return of genial weather fills the country with keep,

and sends up prices.   In the meantime we are put to shifts of various kinds.   We, for example, are obliged to feed grass which ought to be reserved for mowing.   We are compelled to stock our spring fodder crops before they reach perfection, and our rye, winter barley, trifolium and vetches are in turn invaded before they have attained half their growth.   Over-stocking empties the exchequer, by reason of the hay and cake bills which accrue from it, and the farmer is in a state of chronic discontent and anxiety, waiting for growing weather. The evils of overstocking are not confined to a protracted shortness of victuals.   They are seen in an increased mortality, a prevalence of scour, and a dryness of coat, and want of bloom among lambs.   In long-woolled flocks it results, at length, in loss of size and frame in the ewe flock.   Such are the evils of overstocking with sheep.

We should, therefore, endeavour to keep within bounds as to the number of sheep, and at the same time arrange our cropping so as to give an abundance of victuals throughout the entire year.

The foods principally relied upon during the winter are turnips, swedes, mangels, clover and meadow hay, peas-straw, oat and barley straw.   These may be called the natural foods produced on the farm, and a few words upon each of them may not be out of place.

Turnips

are usually stated to be insufficient for fatting purposes.   It has, in fact, been declared that sheep cannot be fattened on turnips alone.   It must be allowed that the experiment is never strictly made in practice, because some sort of dry food is always given. I have, however, seen sheep do remarkably well upon turnips, and Mr. Clement Cadle, of Gloucester, who I regarded as an eminently practical authority on farming, told me that he had seen them fatted on turnips and little else.   Turnips and straw are capable of keeping sheep in good condition, and so far as nutritive properties are concerned, they are sufficient. Too little regard is paid by purely scientific authorities to the great variation of quality in turnips according to the ground upon which they are grown. To them a turnip is a root containing 92 per cent. of water, and this is enough to condemn it. Turnips grown on land of good quality are superior to those grown on weak, poor soils, as is evidenced by the quicker progress made by sheep placed on good land. White turnips, although actually inferior to swedes as a food, according to analysis, are superior to them up to January 1st, or even later. They are less trying to the digestion, and their consumption forms an excellent introduction to the harder winter feeding which begins with the new year. Store sheep will thrive well upon turnips if they have access to oat straw or a little long hay or hay chaff. Ewes should be allowed a more liberal supply of dry food, so as to induce them to eat more sparingly of the succulent turnip. Fatting sheep must have hay and cake or corn, and it will be found that the diminution in the amount of turnips eaten will be at the rate of 12 lbs. of turnips for each 1 lb. of cake supplied.

When turnips are being fed off the ground it is a good plan to pull them up with a pecker, and allow them to lie on the ground a day or two before they are folded over. This dries them a little, and renders them slightly pined and soft, and thereby heartier and less watery.

SWEDES

contain somewhat less moisture than turnips, and considerably more sugar. They are a stronger food, better adapted for fatting sheep than for ewes either before or after lambing. They should be reserved for the later periods of winter feeding from January to May, after which they rapidly lose in quality. Swedes, because they are harder than white turnips, and because they are used when tegs are losing their central teeth, should be cut into troughs by means of a Gardiner's turnip cutter.

MANGELS.

My attention has lately been called to the fact that mangels cannot be safely given in quantity to wether sheep. They appear to have an injurious effect upon the bladder and urinary organs, and in many parts of the country flock-masters dare not use them for wethers, although they may be given to female sheep. On this point of practice I may say that mangel is relied upon as a change of food for ram lambs in the great sheep-breeding district of Wilts and South Hants. The roots are either cut into troughs or thrown about upon vetches or rape as a change, but the lambs have at that season so large a variety of food in the shape of green fodder, that no evil consequences follow. Ram breeders would find it difficult to do without mangels in the hot months, and shepherds may be seen in our showyards slicing mangels and giving them with cabbage and rape to rams.

HAY.

It would be thought extravagant in the north of England to give hay to sheep. There a smaller amount is made, and that chiefly for horses. Experiments upon the powers of digestion of our domestic animals have proved favourable to the use of hay for sheep, and in the great sheep-farming districts of the chalk and oolites, shepherds would feel sadly aggrieved if hay were withheld. Horses, and even cows, are made to live on straw before the flock, which becomes the great consumer of hay. The best descriptions for sheep are clover and sainfoin hay, and the best method of feeding is in the form of chaff. It is cut up by means of powerful three-knife chaff-cutters, worked by steam, and furnished with screens for separating the dust. In these machines the product is delivered into quarter bags, which are carted to the field and emptied into troughs. The chaff-cutter is also useful because it enables the farmer to mix a certain quantity of straw with the hay, but such a proceeding is not agreeable to the shepherds, who, in spite of all that is written in books on the subject, believe in pure and unadulterated hay. We must, in fact, regard a good stock of hay as one of the greatest boons and best safeguards against those evils to which a ewe flock is exposed during winter. It is a warm and comforting food, and mixes well with turnips. It is this combination of dry, sweet hay with watery turnips which, more than anything else, will preserve the high position which this food holds among practical men. Ensilage would not meet the case nearly so well, and, useful as silage undoubtedly is for dairy stock and fatting bullocks, it is not likely to oust hay from the sheepfold.

STRAW.

Sheep should be allowed access to straw, and in hard weather will eat a good deal of it. Like hay and other dry food it helps to prevent over-indulgence in roots. Straw is, however, scarcely nutritious enough for sheep, and is better bestowed on cattle in sheds and stalls, when it can be moistened with water or with mucilage made from linseed. The system of pulping roots and mixing them with straw chaff also is excellent for cattle, but less applicable for sheep in the open field. Pea haulm is richer than the straw of cereals, and sheep are very partial to it. The pea crop might be cultivated more widely with advantage by sheep farmers, as both the grain and the straw would be found exceedingly valuable foods. Next to pea straw, oat straw is the most useful, and barley and wheat straw can scarcely be recommended in quantity for the flock.

COTTON CAKE.

Shepherds look askance at " that yellow cake." Whether from custom, prejudice, or real experience, they prefer best linseed cake, and surely there is much to be said in favour of this opinion. And yet excellent results have been obtained with cotton cake both in teg and ewe feeding. For lambs it is not suitable any more than it is for calves. Rough or undecorticated cake may be given to ewe tegs without fear, and a mixture of cotton and linseed cakes to fattening sheep. Ewes also eat it with relish and do well upon it, but here its use ends. The case of cotton *v.* linseed cake for manurial purposes, although at one time loudly proclaimed by agricultural chemists, has not been distinctly proved in practice, and the merits of the rival cakes will probably continue to be tested by feeding results rather than by manurial effects.

BEANS AND PEAS.

Linseed cake and old beans are the staple concentrated purchased foods of successful sheep feeders. White peas are in high repute for lambs ; but as these animals become older, opinion inclines towards beans. Pea chaff (husks) is also largely used in spite of a somewhat unsatisfactory analysis. Malt coombs (culms) is another excellent food, especially for mixing with inferior hay chaff.

THE USE OF CAKE AND CORN.

Profitable sheep farming is inconsistent with too lavish a use of cake and corn. Fatting tegs may be allowed from ¼ lb. to 1 lb., or even 1½ lb. of cake and corn per day, and in extreme cases 2 lbs. may be given. The quantity should be gradually increased from, let us say, 2 oz. to 2 lbs. When a large flock of even the highest quality is kept, the use of cake and corn should be restricted to the time between yeaning (lambing) and weaning, or at latest to shearing. Many farmers continue to give a little cake to ewes after weaning up to shearing, because the wool then comes off easier and better. The breeding flock from June to the following Christmas should live off the natural produce of the farm if they are to prove a profitable stock. The case of the lambs is different, and much depends upon their destination. Ram lambs should be pushed on as fast as possible, whereas ewe lambs should have little or no cake, the quantity depending upon the object which the breeder has in view. Store lambs will do very well on summer fodder without extra assistance. In making these remarks I have in view flocks in enclosed districts, and not hill flocks, which are so little acquainted with cake and hay-chaff, sliced turnips, and other artificial modes of feeding, that they have been known to stampede at the very sight of them and their accessories.

Open Grazing and Close Folding.

A general survey of the methods of grazing sheep during summer shows two principles in ordinary use—open grazing and close folding. The first of these systems prevails on all sheep walks and in a number of enclosed districts. Perhaps the most ordinary idea of feeding sheep in summer is to turn them into fields and allow them to roam and select their own food. Ewes with their young lambs are put into a fresh piece of clover and rye-grass, and, where the management is liberal, a few troughs are to be seen in which oats or a little cake are supplied daily. The management consists in inspection from time to time, and the chief evils to be guarded against are scour, flies, sore heads, breaking through fences, and other casualties which must be seen to. These are so numerous that a farmer seldom walks round among his sheep without feeling pleased that he has paid them a visit. At one time he finds a ewe on her back in a furrow, at another a lamb entangled in brambles or hung up in a gate or sheep-net, or he may just arrive in time to prevent the whole flock from wandering through a gap into a field of standing corn. It would, therefore, be untrue to say that this system involves no care or trouble, but it is nevertheless comparatively simple. The general result of this system of open grazing is satisfactory, but it cannot yield extraordinary results, unless indeed at a great sacrifice of area and of food. When the master judges that the pasture is becoming bare, he shifts his flock to another, and allows the first to regain its freshness, when it may again be stocked. This system of free and open grazing is constantly to be observed in travelling through the country. The sheep are dotted over the pastures, or, as the express train dashes past them, they are seen scampering away from the fences in evident perturbation of mind and body. The system is not only generally practised, but has its advantages when used as an alternative to close folding, even in the case of sheep intended for exhibition, or for the maximum development of the animals.

Open Grazing for Show Sheep.

Show sheep require room, and even in districts where close folding is the rule, the favourites are allowed to wander. How often have we heard the advice given to allow such sheep to " lie about " on a piece of good sainfoin, or to run forward unchecked while the remainder of the flock follows restricted between nets or flakes. By this system, if overstocking is avoided, the best choice of food is given, and it may be recommended during at least a portion of the day. Our Down breeds are well known to be patient of the fold, and to be capable of being placed thicker on the ground than are the Longwoolled races; but even Down sheep are the better of a free hand in the selection of herbage. Perhaps the best plan is to combine the advantages of both systems in such cases, and while allowing liberty at one part of the day, to confine to a fold during another period.

In hot weather sheep often seem to flag and sicken, and refuse to eat during the most oppressive hours, and when this is noticed the shepherd should draw off his sheep to the open and breezy down, where they may spread themselves, and they will then be seen to regain their appetites and pick about among the young grasses and clovers. They will then return to a fresh fold on which they will feed heartily towards sunset, when they should be put back upon the old fold and allowed to rest.

Change of Food

is one of the secrets of successful sheep farming. Wherever this pursuit is carried out to the best advantage, at least one change is made in their pasturage during each day. It may be from ordinary seeds to down, or from old sainfoin to seeds, or from seeds to permanent pasture. The exercise in moving from one part of the farm to another is beneficial, and the change of food is agreeable and stimulates growth. While, then, we do not object to open grazing as a principle, we deprecate the system of turning sheep into a field and there allowing them to remain until the herbage is exhausted and a change is simply imperative.

Close Folding

is the rule during winter while the flock is upon turnips, but is rather the exception during summer. Over whole counties the system is unknown, while in others it is ordinarily followed. It is most characteristic of those districts in which Down sheep are kept, and where the arable land is adapted for the growth of fodder crops. As already explained, close folding should not be pushed to an extreme, but should be alternated with a run in open fields.

The advantages of close folding, when judiciously practised, are very great. A fresh bite is insured, and this is especially the case when two pitches out of the food are made daily. The sheep may be allowed upon a fresh " break " until they begin to lie down to rest, and they can then be shut back upon the old fold. After a while they are again allowed access to the new portion, which then becomes the resting-place, and a fresh fold is struck out in the luxuriant clover or rape, or whatever the feeding crop happens to be. Thus the animals are taken gradually and systematically over the entire area, and the food is always fresh and untrampled.

Sale sheep are not asked to clear up the whole mass of food, and the close-folding system allows of a perfect regulation in feeding. What the sale sheep leave is eaten up by the stock ewes who follow their more favoured offspring, and gnaw down the fold to the bare ground. A regular system is often followed, by which perfect economy of food is insured. The most forward wether lambs are allowed the first pick, and may be pushed forward without regard to waste. They are allowed to pick the choicest morsels, to bite off the tops of rape or clover, and to eat the hearts out of white cabbage. A second and less forward lot may follow, and still further demolish the feast, and, finally, the old ewes are admitted to clean up and make close work. This completely does away with the charge of wastefulness, and is less expensive than moving and throwing the cut herbage into racks. It is also better than the system, which has its advocates, of constantly shifting hurdles and allowing the sheep to feed with their heads through vertical bars. This system we cannot recommend. We have seen hurdles specially constructed for the purpose. They are made in the form of an equal limbed cross, two legs of which rest upon the ground, and by turning them over, a fresh yard of fodder comes within reach of the hungry sheep. We object to this proposal on the grounds of trouble (expense), of insecurity, and of insufficiency of food. Wherever a mixed flock is kept the different sections may be relied upon to make perfectly clean work and prevent all waste.

The Lamb Hurdle

or lamb creep is a useful and necessary institution when close folding is the rule. By its means lambs are saved the hardship of close penning upon a stale fold. Lambs must have change and freedom, and both are secured by the lamb creep.

They can run forward to crop the choicest and sweetest herbage, and return to their mothers for milk. Neither is it necessary to allow lambs too large an area, as a large outer fold fenced with hurdles, say, of a quarter acre in extent, is sufficient at once, and as the ewes advance, the outpost devoted to the lambs is also pushed forward.

Close folding allows of the greatest amount of variety in feeding. The sheep may, for example, return to a fold of vetches the last thing at night and feed upon a fresh bite in the early morning. A good shepherd should be up early, and if he can give his charge a feed at five o'clock they eat heartily before the sun is powerful, and will take their corn at nine o'clock with an appetite. He will then grind up some mangel or white cabbage, and toward mid-day transfer the sheep to a fold of rape, which ought to be provided near. At four o'clock he will give them a run upon clover aftermath or sainfoin, and towards evening he will take them back again to their lair upon the vetches. Thus a happy day is spent, full of variety, and this is the way in which lamb prodigies are reared and astonish the eyes of ordinary farmers from districts where sheep grazing means turning out into a grass field. The difference in value in favour of the close-folding system, as just described, will be about 20s. a head on lambs in September. What would have been fair stock lambs, worth 25s. to 30s. each, are developed into animals worth 45s. to 50s. without difficulty. As to cake, it is no doubt an important consideration, but it pays to give it. When the best cake is under 1d. per lb., half a pound on an average over the whole period of summer grazing is only 3d. per week, some of which must be allowed to remain in the land. If we allow, as we think we may, half of the cake to be left in manurial value, the charge upon the lambs for cake is reduced to 1½d. per week, or during a season of twenty-four weeks to 3s. per head. If, however, as might be preferred, " every tub should stand upon its own bottom," and " every herring hang by its own tail," then we think the direct expenditure of 6s. per head upon caking lambs will prove remunerative, and the benefit to the land will come in as a pleasant reflection which will no doubt in time be realised.

FODDER CROPS FOR SUMMER FEEDING.

The high estimate of the value of fodder crops has of late years been raised by researches as to the waste of fertilisers through drainage. It has been shown that fodder crops prevent waste by appropriating fertilising matter, which, but for them, would flow into drains, or into the inaccessible parts of the subsoil. Sheep play an important part in conjunction with fodder crops in this important matter. The crop, in the first case, is a collector of plant food from the air, and the soil, including the subsoil; and, having fixed certain important constituents, the sheep convert them into a marketable commodity, returning a large proportion of the food constituents to the soil, thus enriching the surface. Sheep have been very properly spoken of as an exhausting stock to the land, but while this is the fact, the reverse is also true; for sheep may be used to improve land, more perhaps than any other agency. We know that in Kent and other counties a ewe flock is looked upon as scourging, chiefly on account of the drain it entails, upon the land, of phosphates. In Sutherlandshire, sheep grazing has impoverished large tracts of mountain pasture. Such facts only teach the necessity of carrying on sheep-farming upon rational principles, and especially show the importance of feeding with artificial foods, as well as with the natural herbage of the farm. Bearing this in mind, we may safely say that the more sheep a farm carries (within certain limits) the better, for then the farmer is compelled to assist his flock with artificial foods.

The growth of fodder crops, and their consumption with cake and corn, are the best means of increasing a sheep stock. The proper conditions must, however, be observed, for it will not answer any good purpose to attempt to force a system on unwilling land or under unfavourable skies. In our southern counties, and on light-topped soils, the most suitable circumstances are found for this system of management. Here the fodder crop may be followed with a full root crop, and then the land may be made to support sheep twice—first in summer, and again in winter. The most suitable crops for a system of summer fodder-growing are, first,

The Winter Cereal Fodder Crops,

such as rye, winter barley, and winter oats. Of these, the first is the earliest. Rye yields an excellent sheep food. It may be sown any time from midsummer to September. When a turnip crop has failed, and repeated re-sowing has not succeeded, rye may be drilled. Such early-sown rye will furnish a fold for sheep in the autumn, and again in the spring, after which the fold may be broken up and roots planted. More ordinarily rye is sown after a corn crop, in September, and gives a spring feed, or catch crop, before roots are sown. Rye is a wholesome food for sheep as it is slightly binding in its nature, and is thus well adapted for feeding with more laxative foods, such as water-meadow grass or turnip greens. It must be fed young, and before it throws up the seed stem. It leaves the land in good time and in good condition for the earliest root crops, such as mangel wurzel and early rape or turnips.

Winter Barley

may be fed with rye. It comes into use slightly later in the season, and is therefore less adapted for following with early root crops.

Winter Oats

are still later, and assist to prolong the period in which these cereal fodder crops may be employed. By the use of these three crops there is no difficulty in producing fodder during about two months of the spring.

12

TRIFOLIUM.

Sheep are very fond of this plant, and thrive well upon it until it begins to grow old and hard. It is most useful at the beginning of summer and precedes

VETCHES,

which must be considered as the mainstay of the flock during the hot months of June and July. While vetches are in use the earliest-sown

RAPE

becomes fit for stocking, and shortly afterwards cabbages may also be relied upon. Vetches, rape, and cabbage assist to tide over the latter weeks of summer before the earliest-sown turnips arrive at maturity.

Besides these crops, sheep will probably have access to young seeds and to aftermath clover, and, if possible, a supply of mangel wurzel should be also retained until the turnip season comes round again.

The various fodder crops named are better fed in combination than separately. Thus rye ought to be fed with water meadows and turnip greens, as well as with swedes and mangel. Winter barley should be fed with winter rye, mangel, and trifolium or winter oats. Trifolium should be fed with vetches; vetches with rape and clover aftermath; rape with cabbages and clover aftermath. When to this variety are added sliced mangel, linseed cake, and beans or peas, a varied food is constantly provided, and the things can hardly help thriving.

As the acreage is cleared the ploughs follow, and the land is quickly covered by a thriving turnip crop, which supports the flock during the winter. The amount of sheep stock which may be maintained during the summer months under a system of fodder crops is very great. It may be best gauged

by the number of ewes which are maintained, and this generally amounts to one to the acre. Taking a 500-acre farm we
should, in the first place, expect to find a dairy of twenty-five
cows, together with young stock, bringing the entire head of
horned stock up to fifty. There will also be the usual number
of working horses, with pigs and poultry. The total sheep
stock on such a farm would include during the winter—*i.e.*,
from October to January—500 ewes, about a dozen rams, and
from 150 to 180 ewe tegs. When lambing is completed, the
sheep stock will have increased to the following numbers,
which may be taken as approximately true of many farms:—

480 ewes
160 ewe tegs
12 rams
500 lambs
———
Total sheep stock 1,152

This stock is maintained throughout spring and summer,
and only begins to be drafted in early autumn. The cull ewes
are the first to go, and these make room for the ewe tegs.
The best wether or ram lambs and some of the ewe lambs
follow, until by October the head of stock is reduced to the
old limits. In some cases 1,200 or 1,300 sheep are maintained
during the summer upon 500 acres, and, as the lambs grow
rapidly, they are equivalent to the sheep of many districts
in size and appetite. Meanwhile the corn area is fairly kept
up, so that upon such a farm of 500 acres from 160 to 200
acres of corn will be grown. No description of farming is
more likely to prove profitable at any time when the price of
sheep is maintained, and when good barley can also be
grown.

CHAPTER XVIII.

EXHIBITION SHEEP.

THE production of the splendid specimens of sheep which form a principal attraction at all agricultural shows takes us out of the ordinary routine of farm management, and introduces us to one of the mysteries of the art of stock-keeping. The contrast between a pen of wethers at Islington, or a pen of ewes at any great show of breeding stock, and ordinary wethers or ewes as seen on the farm or at a fair, is very striking indeed. How such a result is obtained must indeed appear a curious question to the uninitiated, and it is not too much to say that it could not be achieved at all without a prior outlay of many hundreds of pounds, and an application of skill which could not be emulated by any person outs'de the mystery. We do not say that it is impossible for a man to forthwith enter the arena, or that if enough money is expended he might not at once take his place among the aristocracy of the breeding fraternity. But we do say that his object could only be secured by a heavy outlay, and the purchasing of the results of years, or even generations of patient workers. A first-prize pen at a national show in a country such as England means a great deal. It means a large flock from which to choose, careful breeding for years, and skill in feeding and in selecting the animals sent up. The success could not be achieved without cordial co-operation between master and shepherd, both fulfilling their part in securing the result; and hence, without various qualifications, no outsider could venture with the least prospect of success to compete against those who hold the key of the position. A glance at the requirements of success in exhibiting sheep will readily show that high-class stock breeding is in some respects a close profession, and that not from artificial, but natural barriers. Capital, skill, taste, time, and a suitable situation are all needed, and this combination is rarely to be found in the possession of one individual, and hence we may somewhat selfishly hope that the breeding of really first-rate animals is not likely to be quickly overdone. There is undoubtedly room for many more breeders, and we have never feared injury from competition of this nature. English live stock is in demand over the whole civilised world, and, besides this, breeders are excellent customers to each other.

In making a few remarks upon exhibition sheep, we would say that the importance of the animals being thoroughly well bred is paramount. As well enter a half-bred horse for the Derby as a badly-bred sheep for the highest honours. Neither is it sufficient that the animals should be well-bred on one side, but it must be from a pure-bred ewe and by a pure-bred ram, if it has to win by anything but a fluke.

It is, in fact, but lost labour to train under-bred animals for a show, and hence it is a fundamental fact that the foundation must be laid, either by years of patient breeding, or by spirited buying, in order to start as a fully equipped breeder.

Having, in the acquisition of a good flock, secured the sheep, the next point of importance is the land. This may be regarded as rather a matter of good fortune than of choice. Some farms produce a quality of fleece, a growth of bone, and a mellowness of flesh which cannot be attained on other situations. The occupier sits at a natural advantage with regard to the holders of weaker soils which never can produce the requisite merit. A large farm is better than a small one, not only because the variety of soils is greater, but because there is a greater choice in selecting suitable animals.

Early lambing is an important item in success, but this may

easily be pushed to an extreme. Lambs born too early become coarse and pass their best before they are required for exhibition. Young lambs always have a certain charm in coat and style for good judges, and hence a great deal of judgment is required with regard to the period at which a lamb should be born, or rather with regard to the age of the lamb selected for training. Size is far from everything, and we have seen it passed by for quality on many occasions. Hence an exhibitor may well pause before choosing his exhibits on account of their great proportions.

Preliminaries having been settled, it will be necessary to take out a much larger number of individuals than will actually be required. There are many contingencies to be thought of. A sheep may die or go wrong. He may develop into a coarse brute, or if he goes on right it may be found exceedingly difficult to match him with suitable companions. Thus a pen of three, a pen of five or of ten matchy lambs are not obtained except out of a considerable number, all of whom must be treated alike. This is one of the most serious matters connected with showing. So serious, indeed, is it, that many of our best breeders decline the showyard altogether. They will not force forward twenty of their best ewe lambs in order to pick out three or five for show. The game they think not worth the candle, as the risk of spoiling their best females for breeding is too considerable. The expense is also worth attention, and is not measured by the pen exhibited, but by the total number got ready for show.

As to the best method of getting up a lot of sheep for show, there is the choice between house and open-air feeding. Sheep are not so happy under cover as in the open air, and we have heard the opinion expressed again and again that an open-air life is the best even for show sheep. Any judge can at once tell a shed-fed sheep from his wool. Plenty of room is also a point, and many prizes have been won by sheep which have been allowed to run forward in front of their fellows and pick the primest clover, rape, and cabbage.

Upon the artificial foods it is not necessary to dilate, except in so far as to say that 'sheep of this description should be allowed a plentiful supply of the best that money can purchase. A constant variety in natural foods, and a liberal quantity of the best linseed cake and old beans fairly indicate the food, but who can describe the many minor points as to early and late feeding, frequency of meals, and methods of tempting the unwilling appetite, and coaxing the animals to grow ? These belong to the art of shepherding, and are of vital consequence. A master might as well try to take prizes without sheep as without a shepherd, and it would not be possible to commit all the store of knowledge possessed by a competent shepherd to paper. Neither possible nor yet desirable ; and if it could be done, the written directions would not ensure the same success in other hands. First-rate shepherds are not so uncommon as they are difficult to find, because they are not given to changing their situations often. A pleasant feature of sheep-farming is that mutual regard of master and shepherd, both men appreciating each other's value. Training is carried on with some little affec-tation of secrecy, and much undertoned and almost whispered consultation. The attention is constant and the daily care extraordinary. The trimming of show sheep is a matter of importance. There are those who object to trimming, but it is impossible to show sheep in the natural unkempt and rough state. It is really cruel to ask a breeder to exhibit his sheep in a great show, before ladies and gentlemen, with-out dressing them. What would a horse-breeder say to a regulation insisting that his hunter or his thoroughbred should appear ungroomed and rough, with long tail and uncombed mane ? A sheep-breeder has similar feelings, and similar failings. Besides, the public like to see animals well turned out of hand, and even the pigs appear with their hair curled and oiled, and their skins blooming as if they had been immersed in a bath composed with toilet vinegar. Trimming may be overdone, or unfairly done, but to the legitimate use of the art there can be no objection.   The methods vary with every breed.   The Leicester appears, like the parson, all shaven and shorn.   The Lincoln is smeared over with some mysterious unguent, which makes the hands feel very disagreeable if they are allowed to touch the fleece.   The Cotswold comes out curly in coat, white, and redolent of soap and water.   The Southdown appears as like a plum as a sheep can possibly be made, and bears evidence of the shears over his entire carcase.   A very smug gentleman indeed is the Southdown when in his war paint.   Trimming is carried to the greatest perfection in the Down races, and they certainly reward the artists who have accomplished their tasks so deftly.   How it is done, when it is done, and how often it is done are topics which for the present we need not now enter upon.   Perhaps in a future work we may be able to give full directions as to how to get up a sheep for show, but it must be allowed that there is something almost traitorous in letting out trade secrets.

CHAPTER XIX.

THE FUTURE OF SHEEP FARMING.

If any kind of farming may be expected to pay, even in the present times of depression, it should be sheep farming. In spite of New Zealand and Australian mutton, the home-grown commodity keeps up its value, while only a short time ago store sheep were at a premium, which was rather disconcerting to the grazier. Lamb buyers ought, however, to remember that they also have had their turn in the low prices which formerly prevailed, and previous to that date, when winter grazing must have paid particularly well. The price of sheep no doubt fluctuates, but it does so with the home demand for stock, and the home supply of keep. If keep is scarce, down comes the value of sheep, but with a return of genial showers and quick growth, the price rebounds with wonderful elasticity. How long this will continue, or whether it will always continue, are difficult questions. Professor Wallace says a day of tribulation is coming, but there is comfort in the thought that Australia and New Zealand are not newly discovered countries, and that the same report has been used to alarm us for twenty years, and yet we find mutton selling, as carcases, at 8d. per lb.

The great safeguards against depreciation in the value of sheep appear to be: (1) Their diminished number in Europe, as well as in England, which, although recently checked, is distinct enough to those who study statistics of live stock

Neither do sheep increase rapidly in America, except in the extreme south of the great southern continent.

(2) It is, indeed, with Australia, New Zealand, and the Argentine Republic, that we have chiefly to reckon; and it is also with the Merino breed, which is not a flesh but a wool producing race. The question is important as to how far the Merino can compete with English flesh sheep in our own markets. The fibre is short and the meat is described as delicious, but this is not everything. Weight and rapidity of maturation are more important considerations, and it does not appear in the least probable that the light-fleshed, slow-growing Merinoes are going to oust our own Southdowns from their proud position. English mutton may not be so short in grain or tender as Australian Merino mutton, but it is juicier, richer, better coloured, and the fat is more delicate, and better mixed with the lean meat. It seems scarcely probable that the dry morsel which thinly covers the bones of a Merino can ever rival the succulent and luscious flesh of an English-bred sheep.

(3) Encouragement for the English breeder also is to be found in the same increase of foreign competition in the matter of mutton. If Australian and New Zealand farmers are going to supply England, they must become buyers of English sires and dams, and send their agents to our shores to attend fairs and sales. When we reflect upon the immense areas and the vast flocks of Australasia it is evident that this demand for English blood is practically boundless. As yet, our Colonial friends have chiefly been engaged in buying Southdowns, Lincolns, and Shropshires; but the time will come when they will import the Oxford, the Hampshire, and the Suffolk as freely, and English sheep-breeders will be encouraged to make still greater efforts towards perfection.

(4) The rapid increase of the human family is an important factor in the question of the future price of food. To those who have passed the period of extreme youth, the flight of ten or twenty years must appear as a comparatively short period.

Historically, it undoubtedly is so, and yet the increase of mouths calling for food in such periods is extraordinary to think of. It is not too much to say that in ten years another London is added to our own home consumption! The population of the United States doubles itself in twenty-five years, and the rapid increase of the population of Australia and New Zealand will probably in time require a much larger proportion of their produce than at present. With regard to the United States, we shall have another China in one hundred years, at the present rate of increase, and the question is rather how the human family is to be supported in another century than one of a plethora of food production.

(5) The upward tendency of wages, and the more uniform distribution of wealth, is a feature of the present day. The labouring population is no longer contented with bread, but wants meat, and, what is more, gets it. There is no doubt that comfort and plenty have descended to a strata which formerly lived in comparative squalor and want. Just as wheaten bread has taken the place of rye and barley meal, so beef and mutton will more and more take the place of pork and cheese, and thus a market will be found for all the butcher's meat that can possibly be produced. A drop of 1d. a pound in the price of meat must always call in tens of thousands of buyers to check any further decline in value.

Taking all these considerations into account, it seems then hardly probable that any sudden increase in importations of frozen mutton will overturn the home manufacture.

A few years ago Mr. Finlay Dun returned from a trip to America with glowing accounts of the endless supplies of beef which were about to inundate us. Later, his countryman, Professor Wallace, returned from a tour in Australia, and prophesied evil things of the sheep trade. It is more than probable that his gloomy forebodings will also remain unfulfilled. No one ought to be more pleased than Professor Wallace, if I am right.

The immediate future of the English sheep-farmer is, it is

true, gloomy. He is depressed by low prices, and holds only a small stock of good hay, while the root crop is fairly abundant. Corn is cheap, and cake to some extent participates. He, however, cannot fear over-production, for he is millions of heads below the sheep census of a generation since. Rents have come down enormously. Two farms which I have in view, formerly let at £1,400 and £1,200, have been re-let respectively at £500 and £400, and such cases are no doubt common.

It is probable that the present tenants of these farms are much better off than if they held land for nothing in Australia. At what price per acre, I would ask, ought we to value the mere fact of being in England? Surely it is worth several shillings per acre to be within eighty miles of London, or fifty miles of Birmingham or Manchester. It would be difficult to say how much; but surely land rented at 10s. per acre for sheep-farming in England must, in hundreds of cases, be better than free land at the Antipodes. Climate alone is a consideration. An American once said of our climate—"that we are always grumbling about it, but it was the only thing we ought to be proud of." The remark was sufficiently caustic, but from a grazier's point of view there is much truth in it.

I am sorry to hear loud complaints from the North of Scotland as to this branch of pastoral industry. There is, however, a great lesson contained in this distress, as it appears to be in a measure due to outraging the inexorable laws of nature. Generation after generation, lambs have been reared and removed from the great grazing grounds of Sutherland. Bones have been removed, and wool, and with them phosphates and potash, and the pastures have become seriously injured. The area is too large, and the depletion has been carried out too long, for individuals to meet this call, and the dilemma is a serious one, indeed. No one can lay this grievance to the door of " the bad times "—it has been caused by wilful, albeit ignorant, waste, and, even when the times are favourable for sheep-breeders, the men who occupy these exhausted soils cannot take advantage of them. Under such circumstances complaints are only to be expected, but they are happily exceptional, and certainly are not likely to be repeated by the many sheep farmers who duly keep up the condition of their land by a rational system of management.

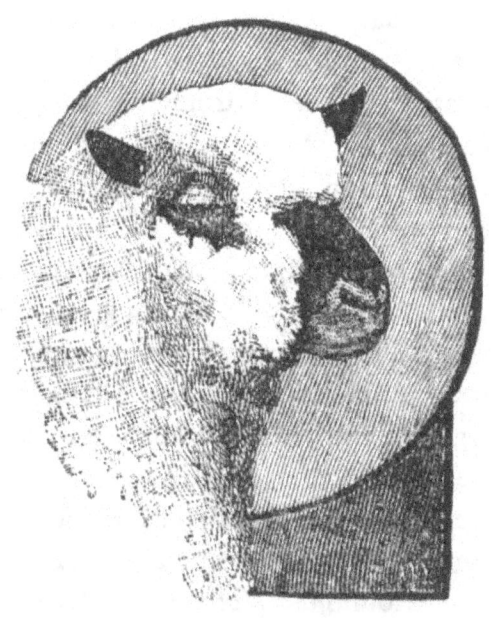

CHAPTER XX.

SHEEP FARMING IN SUTHERLANDSHIRE.

I HAVE so frequently had occasion to speak of the management of sheep on the enclosed grounds of the English shires
that it is now necessary to turn our attention for a short time
to the comparatively wild life of the Highlands of Scotland.
The ordinary notion of sheep-farming in England is naturally
limited to holdings of moderate size. Five hundred acres
is a fair-sized farm, and a thousand acres strike us as large.
Ewes, with corresponding numbers of lambs and tegs, careful attention, regular turniping, and allowances of cake and
corn are associated with arable farming, and especially with
barley-growing. Such *intensive* farming supplies plenty of
food for remark and discussion. The "management" is constantly varying and very complicated, the changes of food
frequent and elaborate, the expenses per acre, or per head,
heavy. All this is entirely different on a Highland sheep-
walk, and we may almost say that the picture is the reverse of
what we often see around us. The term *extensive* is more appropriate. It is not a question of how many sheep to the acre,
but how many acres to the sheep. We deal not with hundreds,
but with thousands of acres ; not with rounded chalk hills or
wolds of oolite, but with lofty mountains, whose tops rake the
clouds and bring down their treasures in torrents ; snow-capped
peaks and distant blue vistas of rugged scenery ; such pictures
as Landseer loved to paint, with the stag and the deer-stalker
in the fore-ground, or the tartan-clad shepherd with his faithful

collies. The whole surroundings are changed, and the dialect in which the master talks to his herd is almost unintelligible to a southern ear.

" The traveller who crosses the mountains to the north-west coast from Kildonan," says Mr. James Macdonald, " enters the Sutherland portion of the parish of Reay. The principal holding in this district is the fine farm of Bighouse, extending to about 60,000 acres, and rented at £1,262 by Mr. Robert Paterson, of Birthwood, Biggar, Lanarkshire." This farm is about double the size of the small county of Clackmannan, and a great deal larger than Kinross. If Rutland were so divided there would be room for one and a-half such farm. The rent amounts to fivepence per acre, over all. Along the banks of the river and burns there is a good deal of green land, but the greater part is black and mossy. Rhifail extends to about 30,000 acres, and is rented at about £900 a year, or sevenpence per acre, and consists of mixed hill pasture. Ribigill, the largest farm in the parish of Tongue, and one of the best managed in the county, is 30,000 acres in extent, and is rented at £1,465 a year, or at about one shilling per acre. Melness, the largest farm in the county, and probably in the kingdom, is 70,000 acres in extent, and is rented at £1,257 per annum, or a little over fourpence per acre. The entire stocks of the districts are of Cheviot sheep, and have been so for more than half a century, and during all this time the change in the system of management has been so slight as scarcely to be worthy of notice. The pastoral farms of the county carry what are known as " ewes and wether " flocks, that is, self-supporting flocks that throw into the market every year a crop of cast ewes and of three-year-old wethers. These ewes and wethers are delivered to the buyers direct from the hills, about September 1st. Sutherland is a breeding and rearing, and not a feeding county. Mr. Macdonald tells us that the best idea of the details of management in Sutherland may be obtained by taking the case of a tenant entering a farm at Whitsuntide, the usual time of entry, and by following him through his first

twelve months. He would take over from the way-going
tenant at valuation a flock corresponding to, say, 2,000 ewes.
In an ordinary year from seventy-five to eighty-five lambs are
nursed for every 100 ewes tupped, so that with the 2,000 ewes
there would be at least 1,500 lambs. The death-rate in
ordinary seasons is from 5 per cent. to 10 per cent. Taking
it at the higher figure, the incoming tenant would find 1,350
one-year-old sheep (say 675 gimmers and 675 dinmonts),
and in round numbers 610 two-year-old wethers, and 550
three-year-old wethers. He thus would commence with a
grand total composed as follows :—

> 2,000 ewes
> 1,500 lambs
> 675 gimmers ⎱ one-year-old sheep
> 675 dinmonts ⎰
> 610 two-year-old wethers
> 550 three-year-old wethers
> ———
> 6,010

Besides two or three "turnip herds" there would be seven
shepherds on the farm, each having under his charge 500 ewes
and their lambs. The ewes would occupy the lower and
greener land, the others the higher and blacker. Clipping
commences with the "eild" sheep about the first week of June
and continues for a month—the smeared sheep having been
washed immediately before. He sells his wool and his three-
year-old wethers, and, perhaps, his "shott" lambs and draft
ewes at Inverness Wool Fair, where he probably also sells his
wool. The weaning of lambs takes place in the last week of
July or first week of August. The worst lambs are sold, and
the remainder are grazed on the better pastures until the first
week in October, when they are sent for wintering either to an
arable farm on the south-east coast, or to Caithness, Ross, or
Moray, or elsewhere.

The hill stock being thus lightened, the wethers occupying
the higher and colder grounds are smeared with a mixture of
tar and butter in equal proportions, with perhaps a little oil and
milk added, the cost being about 8d. to 10d. a head. The

remainder of the stock is dipped with oil and tobacco juice, at a cost of 3d. per head.

The tups are turned out about November 22nd, and taken back about Christmas, from forty to fifty ewes being allotted to each, and a few of the strongest or hardiest gimmers are also tupped. Two or three of the better bred tups are mated with choice lots of ewes in order to breed rams for future use. During the winter the younger sheep are placed in the best ground, and from each " hirsel " weakly animals are drawn, and given hay and oats, if necessary. If the winter is open and the pasture good, none except specially weak sheep require hand feeding ; but if the winter is severe the younger " hirsels " get a little hay, or a small daily feed of turnips, if these are raised on the farm. Hand-feeding is considered a bad practice on hill farms, and is only resorted to when it becomes a necessity. The ewe and wether tegs are wintered separately, the former on grass as much as possible, getting turnips for a month or more in spring, while the wether hoggs are wintered on turnips for about six months, at an expense of from 8s. to 10s. each.

The cotton grass, or mossing, is ready to take them by the end of March, and the hoggs are then dipped and sent on to the cotton grass, which maintains them till the deer-hair and other grasses come up in May. Lambing commences about April 20th. In May the lambs are branded with the farm and hirsel marks, and the males are castrated at the same time.

The above simple account of a year's sheep farming in Sutherlandshire indicates that the number of events of interest are few, and that the minute attention which is daily necessary in an English flock is not required. The sheep, indeed, live an independent and natural existence, exposed to many vicissitudes during the long winter, and enjoying a free and happy life in the summer.

With regard to the number of sheep which a large Highland sheep farm will carry, Sciberscross, which is 39,000 acres in extent, is stocked with 6,000 sheep, and is rented at £1,390. Kinbrace, 30,000 acres in extent, carries about 5,000 sheep, and is rented at £990. As already pointed out, a farm supporting 2,000 ewes would carry a total sheep stock of about 6,000, and might be 30,000 to 40,000 acres in extent. Probably the largest extent of land held at one time in a single tenancy was 150,000 acres under the occupation of Messrs. Donald and William Mackay (father and son). These gentlemen were considered to be the most extensive farmers in the United Kingdom. Although Cheviot sheep are said to be almost universal in Sutherlandshire, there has been a disposition of late years to return to the older Black-faced breed, chiefly on account of the severity of the winters. The deterioration of the pastures is a cause of much anxiety, due apparently to continued grazing without making the necessary return in the form of phosphates. It is also alleged to be due to sporting tenants interfering with the proper burning of the hill ground, and to the greater severity of the seasons during recent years. The simplicity of the feeding, the absence of artificial fertilisers, and the rudimentary character of the buildings, the fixed character of the management, and the vast extent of the holdings are unfavourable to the application of what is generally called science to such a pursuit as this. It bears a strong resemblance to Australian sheep farming, where inspection, change of pasturage, marking, washing, clipping, lambing, and watering constitute the chief occupations of the shepherds. The master is, indeed, more a capitalist with money invested in sheep than a farmer in our sense of the word, and the details of management are chiefly left to the shepherds. No more intelligent and trustworthy class exists than this, and the Collie dogs are only second in importance. The extraordinary sagacity of these dogs has often been illustrated, and the manner in which they watch and marshal the flocks is admirable.

CHAPTER XXI.

A LARGE FLOCK.

A FLOCK of a thousand ewes is unquestionably a valuable property. Such a large flock I have recently inspected, and a few remarks upon what it entails in the way of total sheep stock may be worth perusal. This flock is maintained upon about 1,000 acres of land, varying in quality from rich water meadows and good river-side arable land, to healthy, but poor, down. The ewes are black-faced Hampshires of fair quality, and are mostly used for breeding wether lambs for sale during the autumn fairs. There is also on the same farm a dairy of about seventy cows, which, with heifers, calves, and the sheep, form an abundant and profitable stock. A thousand ewes, as a standing flock, means a large number of sheep during the summer, when the flock is at its maximum number, and is approaching its maximum value. For the latter we must wait another month, when the cull ewes will be brought up into sale condition, and the most forward lots of lambs will be fit for the market.

Where 1,000 stock ewes are kept 1,100 lambs may be reasonably looked for. The number of lambs born will greatly exceed this figure, but deaths are unavoidable. They are, in any case, sure to be sufficiently numerous to keep the number of reared lambs only slightly above the number of ewes turned out in autumn. Barren ewes form an important item, sometimes reaching a proportion of from 2 to 5 per cent. of the breeding flock. No one can lamb down 1,000 ewes without considerable losses from deaths, barren ewes, and abortion, and although twins may abound, yet the general outcome, when all losses are discounted, will be not far from the estimate put forward.

In order to keep up a flock of 1,000 ewes about 360 young ewes will be required for stock every year, and about 330 cull or draft ewes will be available for sale. Such estimates must be approximate only, but if we assume an annual loss of 5 per cent. (a loss which really ought not to be incurred) we should have the actual reduction in numbers through death represented as follows :—

$$
\begin{array}{llllll}
360 \text{ two-year-old ewes, of which about 4 per cent. die} & = & 15 \\
345 \text{ three-year-old} & ,, & ,, & 4 & ,, & ,, & = & 14 \\
331 \text{ four-year-old} & ,, & ,, & 4 & ,, & ,, & = & 13 \\
\end{array}
$$

1,036

This would allow a sufficient number over the 1,000 to stand against casualties during the season.

The flock during April and May, and in fact through summer, will therefore consist of, say, 1,000 ewes, 1,100 lambs, 360 tegs, and probably about ten or twelve rams left over from last autumn. The total number of sheep on the farm would therefore be 2,470, or approaching $2\frac{1}{2}$ per acre.

REALISATION.

A question of some interest very naturally springs out of this statement. What may be expected from a flock of 1,000 ewes, in revenue? This question may be answered either in terms of so much per ewe or so much per acre, and, as 1,000 ewes seem to correspond to 1,000 acres, the two would allow of a similar answer in either case. The amount realised must depend in a great measure upon the quality of the flock. I will take the case of a good ordinary flock, such as the one under consideration. The breed is Hampshire Down, and the locality Wiltshire.

What then is likely to be the proceeds of these 1,000 ewes?

In the first place there is the clip of 1,370 sheep, which may be taken as averaging 4⅞ lbs. each, or slightly under 5 lbs. This equals 6,580 lbs. at 11½d. per lb., or £315 5s. 10d., less draft, &c., £312 10s. 6d. The next item will be the sale of 320 cull and fat ewes, which, if good, we may take at 45s. each early in August. This gives a return of £720. The lambs may fairly be considered as half composed of wethers and half of ewe lambs, of which 190 will be for sale. I estimate the values of the lambs sold in August as follows :—

	£	s.	d.
100 prime wether lambs at 50s.	250	0	0
100 ,, ,, ,, at 45s.	225	0	0
200 ,, ,, ,, at 35s.	350	0	0
150 smaller lambs at 30s.	225	0	0
100 ewe lambs at 40s.	200	0	0
90 ,, at 35s.	157	10	0
	£1,407	10	0
Deduct for losses	45	10	0
Net sales £1,362	0	0

The total proceeds of such a flock would therefore be—

	£	s.	d.
Wool	312	10	6
Ewes	720	10	0
Lambs	1,362	0	0
	£2,395	0	6

There would also be the sale of cull rams, casualty sheep, and skins, which would probably bring up the sum to £2,400 or thereabouts. It will, therefore, be seen that the return in sales from a flock of 1,000 ewes maintained on 1,000 acres would amount, in such a case as this, to over £2 each ewe, or each acre.*

* This estimate was made when sheep ruled high, and prices might easily be 10s. or even more per head below those given in the text. On the other hand we have seen prices higher.

PROFITS.

It is exceedingly difficult to come to a conclusion as to profits. Attempts have been made to reduce this question to one of book-keeping, but such methods are fallacious. It is impossible to say what a flock costs, because the whole business of farming is inextricably interlaced. The grazing of sheep upon clover, for example, is beneficial to succeeding crops, and folding upon turnips is not to be charged to the sheep alone. Cake and corn given to sheep are also partly expended upon them and partly upon the land, and labour is similarly difficult to apportion. However exact a system of book-keeping may be, many of these items could only be assessed by the exercise of judgment, and they would therefore be open to argument, and cease to be of the value which the figures appeared to indicate. Gross revenue can be easily ascertained, and after that no system of books can improve upon careful thinking out of the items of expenditure with a view to forming a judgment. There can, however, be no doubt that sheep are profitable. One of the saddest features connected with the agricultural depression is that many farmers have parted with their sheep, and were unable to participate in the good prices which sheep farmers recently realised. Corn and roots, mutton and residual fertility, are so mixed up in the general conduct of farming business that it is impossible to separate them except by the rough-and-ready system of halving or quartering the expenses. Such guesses, when put down in books, are accurate in appearance, but rest on a weak foundation. I would rather take a general view as to the cost per head of, say, 1 lb. of corn per day at ¾d., and add 4d., 5d., or 6d. for the natural food of the farm, according to its abundance. Add to these items the cost per head of attendance on a basis arrived at on a covering scale. Add to this, losses at 5 per cent. per annum, which would probably not be more than 1s. per week per score, or less than ½d. per sheep per week.

Sheep will, by such a calculation, be found to cost a figure

circulating around 1s. per week, and varying from 6d. to 1s. 3d.,
according to the food given and the age of the animals.   It is
by such a carefully-thought-out estimate, made in conversation,
or in a quiet walk among the sheep, that the farmer can best
and most truly think out the question as to whether his sheep
are paying or not.   It is often considered that in ordinary
circumstances, if a lot of sheep is paying 1s. a week per head,
they are doing well.

Applying this rule to the ewe flock, according to the calculations above made as to gross revenue, they pay about 9¼d.
per week each.   It must, however, be remembered that during
by far the greater portion of the year ewes cost only 3d.,
4d., and 5d. per week, as they seldom receive cake except
when suckling their lambs.   Calculations on such a subject
are, as already mentioned, at best approximative, and it is
only when the margin is considerable that we can absolutely
affirm that the sheep are paying.   It must also be borne in
mind that even a profit of 10 per cent. on the capital invested
in a flock is not a large income, and that a good profit does
not necessarily mean a large sum.

LIBRARY
OF THE
UNIVERSITY
OF CALIFORNIA

COST OF SHEEP KEEPING.

This subject is worthy of further consideration.   So far as
dividing the costs of cultivation in a system of books, we do
not consider it practicable.   It would only be possible in an
elaborate office-kept system of farm accounts, and even then
it would be largely composed of guesses and assumptions.   As
a record of monetary transactions, accounts are of course necessary, and they help to furnish bases for calculations; but
it is a mistake to think that the actual net profits of a sheep
flock, or of a dairy, are ever going to be brought out through
the processes of double entry.   The transactions would need
to pass from side to side of the books like a shuttle in a loom.
Those who understand double entry will be able to follow
me, and will see the insurmountable difficulties in the way

of an accountant endeavouring to show the net profits of a
flock of sheep.   I will endeavour to strengthen this position
by giving an example.

The flock requires turnips, swedes, and mangel, and for
these foods we presume they must be charged.   But before
doing this we must open accounts with turnip, swede, and
mangel, or treat all as "roots," and open a "root account."
The sheep account is thus debtor to the root account, and an
entry is made "Sheep, Dr. to roots."   This is, we will presume, calculated on 3d. or 4d. per week per sheep, but whether
justly or not, no one could very well say.   We must next
make some sort of a guess as to the manurial value of the
roots eaten, for it would be rather unfair to the root crop to
credit it only with 4d. for eatage, and with nothing for its
manurial value.   This would lead to another assumption as
to the proportion of fertilising matter left in the land, and we
should require to debit a manure account or else a field account
with manure received by means of sheep.   This might take
the form of " Gallows Field, Dr. to sheep folding," and would
necessitate either a sheep folding account or an entry on the
credit side of the sheep account.   Or it might be treated as a
transaction between the root account and the field account,
such as "Gallows Field, Dr. to roots feed," and " Roots
credited by Gallows Field."   The debt ought to be carried
on and divided throughout the rotation, involving a large
amount of calculation and a great deal of rough assumption.
When the sheep are debited with cake and corn, they should
also be credited with the unexhausted fertility due to these
feeding stuffs, and the crop benefited would be debited for
the same.   Thus cross entries would multiply, and no one
could say where they had to stop.   Many of them would be
mere bogus entries, for no one could say how far they were
realisable, as much depends upon season, and a heavy dressing of sheep dung might possibly do harm to a barley crop.

An elaborate system of farm accounts to show the net
profits from any department of a farm would therefore be

practically impossible, and, besides, it would take all the farmer's time to keep it.

THE COST OF KEEPING A SHEEP

may, however, be approximately arrived at by a few notes and a little thought, in a fairly satisfactory manner, and of this I can easily give an example. This method also has the advantage of being that which is usually followed by farmers in turning over in their minds whether the system they follow is the best which could be adopted.

We cannot take a better example than the cost of maintaining a ewe, and in doing so we prefer to take the period from weaning to weaning. All such calculations, as already explained, can only be approximative; and, as a principle, we think the estimates should be of a covering character, or should err rather on the side of excess. If a margin for profit is then clearly discernible, we may rest assured that there is a sure profit; whereas, if our returns are below the probable or possible cost, our profit might prove a vanishing one. To estimate the cost of a ewe may not appear a very easy matter, and, in fact, it is not. We propose to treat it in a business-like manner, upon a basis which we know to be correct. For example, keep is a marketable commodity. It can be had at 3d., 4d., 5d., and 6d. per week; we have known it as high as 9d. per sheep per week. Keep can generally be procured at 4d. per head per week even when no cake is fed, and if ¼ lb. of cake is consumed by the sheep on the land, probably 4d. would secure ordinary keep of turnips and hay at any time. The period from weaning to weaning takes up the entire year, which we proceed to divide as follows:—Supposing weaning to take place on June 1st, there is a period of cheap and abundant keep from this date up to October 1st, during which we think ewes ought not to cost in food more than 3d. per head per week. We believe we could secure keep away from home at this rate, and may consider it as leaving a profit to the lessor, and, therefore, liberal. I assume the cost of food from June 1st to October 1st, or for seventeen and a-half weeks, at 3d. From October 1st to February 1st is a time of greater scarcity. The ewes must have hay, turnips and grazing, and I estimate the cost at 4d. per head, believing that in most seasons we could take keep at this figure. On February 1st I shall assume that the ewe produces her lamb, and that she begins to receive a little corn within a few days of that event. I shall, therefore, raise the cost of keep to a figure suitable for couples, and place it at 8d. from yeaning to weaning on June 1st. This figure we may divide, if necessary, into two portions—namely, 5d. for the ewes, and 3d. for the lambs.

Food is, however, not the only item of cost, for we have labour, cost of ram service, risk, actual loss, interest on money, and incidental expenses, such as hurdles, troughs, carting, &c., some of which are exceedingly difficult to estimate. These items, even when estimated on a liberal scale, do not reach a high price per sheep per week. Take labour, for example, as assessed highly at 250 ewes per man at wages equivalent to £1 per week. This is evidently 1d. per week upon 240 ewes, which must be regarded as excessive. As this is a very principal item, we take it at 1d. per week, and we are disposed to put all other expense at another 1d. per week. Losses at 5 per cent. on an average figure of 52s. is 5 per cent. on 1s. per week, which is actually 1-20th part of 1s. per sheep per week, or rather over ½d.

We think we are now in a position to make an estimate as to the cost of keeping a ewe for a year.

	s.	d.
Keep from June 1st to October 1st, 17½ weeks at 3d.	4	4½
„ „ Oct. 1st to February 1st, 17½ weeks at 4d.	5	10
„ „ Feb. 1st to June 1st, 17 weeks at 5d.	7	1
Labour for 52 weeks at 1d.	4	4
Other expenses and risks at 1d. per week, including ram service ...	4	4

£1 5 11½

The cost of keeing a ewe I therefore find to be £1 5s. 11½d. per annum. The lamb has been assumed to cost 3d. per week from birth to weaning time on June 1st, or for seventeen weeks, 4s. 3d., and the cost of the couple is seen to be £1 10s. 2½d.

A fairly kept ewe and her lamb will, therefore, be kept on mixed arable and grass farms, such as abound in the chalk and oolite formations, for about 30s. From this sum it is customary to deduct the fleece of the ewe, which varies with the weight, and cannot be taken at more than 4s. or 5s. Thus the net cost under the system of treatment indicated will be from 26s. to 27s. per annum, and the amount of profit depends upon the price of lambs. This varies exceedingly. In July lambs may be worth 30s., or even 40s., in which case the flock has certainly paid well, but in seasons such as 1893 they are difficult to dispose of, even at prices which it would be misleading to quote.

CHAPTER XXII.

DISEASES OF SHEEP.

[In the section on Management, references will be found to many of the diseases of sheep. As it is not intended that this volume should be a complete veterinary handbook, we shall not now enter into a detailed catalogue of ailments, but the following paragraphs describe a number of the more serious disorders. Several of these notices are from the pen of Professor Axe.]

Sore Mouths and Teats in Lambs and Ewes.

THE article on aphthæ, thrush or canker, by Professor Axe (see page 223), is well worth the careful attention of sheep breeders. There are few breeders but must have suffered at certain periods of the year from the disease which is there described. The consequences are serious, as they entail the loss of many ewes and lambs, and the effects are even more far reaching than this. I remember a severe attack in the spring of 1886, after a scarcity of roots, and a too free use of hay, cake, and other dry foods. The ewes were in good condition, and the lambing season was particularly fortunate and fruitful. A finer lot of lambs, and a seemingly healthier lot of ewes I have seldom seen up to March. It was at this time that some of the lambs were seen to be affected with sore mouths, and simultaneously ewes were noticed to be suffering from a pimpled and crusted condition of the teat. As a precaution, the affected couples were separated from the rest, but the disease spread notwithstanding, and involved a great part of the flock. The spring was cold and late, and harsh east winds prevailed, which further seemed to aggravate the disease. A veterinary sur-geon was called in, and prescribed a number of remedies more or less difficult to carry out over a large flock folded on the open down and exposed to all the rigours of cold nights, morning frosts, and dry east winds. The ewes could not stand to be sucked, and the lambs were perpetually hungry and discontented. Their frequent attempts to suck kept the teats in a wounded, wet and uncomfortable state, while the dry winds caused them to chap. Many lambs were compulsorily weaned, and the task of drawing the diseased and fœtid udders fell to the shepherd, who was well nigh driven to despair. Many ewes and lambs died, while others moved about gaunt and miserable—the ewes tucked up in their bellies, and the lambs long, thin, and dry in their coats. The mischief ex-tended into the autumn, causing many ewes to be drafted before their time; and a further loss was experienced on the sale of the cast ewes on account of many of them being unfit for nursing, and therefore for breeding.

My own case is so well delineated in the article already referred to that the description might be taken from it. I have, therefore, no doubt that we suffered from an attack of aphthæ or thrush. Cure there seemed to be none, and we rather lived through it than mastered it. Warmer weather and plenty of green food, the death of the severest sufferers, and the gradual recovery of others in time, obliterated the ailment, and we had a fairly good lot of lambs to show after all. As I have recently heard of similar outbreaks involving certain flocks, it may be useful to draw attention to the matter, especially as in the article referred to some good directions are given for treatment. A complete change of diet is one of the most obvious and best means for checking the disease. Grass or other green food and a cooling diet would be much better than concentrated and heating foods, such as linseed cake and clover hay. Bran, meadow hay, green rye, water meadows, young seeds, with a few oats, would be likely to change the con-dition of the blood, and produce a healthier and cooler condition

than such heating foods as swedes, peas, and linseed cake.
The udders of affected ewes should be freely lanced and
drawn, and the lambs in such cases must be hand fed. "A
saline aperient, such as a little Epsom salts, combined with
some aromatic, as aniseed or coriander seed, may be given to
the ewes at the outset, and this should be followed by the
administration of a little powdered columba root and salt in
the manger food twice daily. As a dressing for the udder a 5
per cent. solution of carbolic acid applied morning and evening
will assist in healing wounds and dispersing the eruption, or
what is equally efficacious, a weak solution of alum. The
mouths of the lambs should be washed both inside and out
with a solution of chlorate of potash of the strength of 10
grains to the ounce, using just sufficient to moisten the
surface without allowing any to be swallowed. When proud
flesh appears on the lips or gums it should be promptly
touched with lunar caustic, and any teeth which may become
loose should be removed." Youatt recommends that the
mouth should be washed two or three times a week with solution of alum, or diluted tincture of myrrh, and a couple of
ounces of Epsom salts should be administered. He also
describes under the same heading a similar disease which
attacks "lambs oftener than full-grown sheep; and sucking
lambs more frequently than those that are weaned. It is
attributed to various causes, as feeding among the stubbles, or
on stony ground, or the teats of the mother being chapped or
filthy. The application of a little mercurial ointment, very
much lowered with lard, or of the common sulphur ointment
with a twelfth part of mercurial, will speedily effect a cure."
Youatt also points out a curious coincidence between thrush in
the mouth and foot-rot, and suggests that possibly the sheep
may have rubbed the diseased foot with his muzzle or licked it,
and thus communicated the disease to the mouth. What is
known among shepherds as "evil" is no doubt a similar if not
identical disease. It is seen affecting lambs and sheep of
greater age, such as ewes, and appears as an eruption around

the mouth. Lambs eating harsh grass are often affected by it. This disease will be found more fully described at page 223.

LAMENESS.

Foot-rot is said to be a preventable disease. This, although possibly true as an abstract statement, is not always equally practical. For example, ewes heavy in lamb cannot be handled and turned for dressing without danger, and the dressing of their feet is for this reason often postponed. A sheep lame from foot-rot will always be found to have an exuberant growth of horn doubled under the sole of the foot, or splayed outwards in a ragged condition. The sheep should be turned, and every portion of loose and detached horn pared away. All fungoid sproutings should be removed by the knife, and the cutting should be continued as far as the disease has burrowed, or the hoof is detached from the sensitive foot. This we regard as fundamental, and we repeat that every part of the horny portion of the hoof that is in the slightest degree separated from the parts beneath must be cut away with a proper knife for the purpose. Any fungoid granulation must be removed, and then a caustic application must be used. Washing the foot perfectly clean before applying the caustic cannot be objected to, but, in practice, the cleaning is more usually accomplished by direct paring with the knife. As to the " dressing " which may then be applied, it is always of a caustic or strongly astringent character. Butter of antimony applied with a feather is found useful. Also various foot-rot pastes which may be purchased at the druggists, or dry, powdered sulphate of copper, or blue vitriol. Speaking of " butyr of antimony," William Youatt says, " There is no application comparable to this. It is effectual as a superficial caustic; and it so readily combines with the fluids belonging to the part to which it is applied that it quickly becomes diluted and comparatively powerless, and is incapable of producing any deep or corroding mischief." If the foot has been severely stripped of its horn, and especially if a considerable portion of the sole has been removed, it may be necessary to bandage it; but in most cases, after dressing, the sheep is turned away without this safeguard. One or two dressings will usually complete the cure, but in a large flock it is difficult to get rid of it altogether. Prevention is better than cure, and the best safeguard against foot-rot is to dress the feet from time to time so as to prevent the wall of the hoof from growing to such a length as to double under the foot.

This over-shooting of the horn is the cause of pellets of hard earth becoming inserted between it and the sole, and this causes irritation and inflammation. This evil may be prevented by care, and points to the necessity of supervision on the part of the master, and the allowing to the shepherd a sufficient amount of assistance to enable him to give increased attention to the health and comfort of his flock. Foot-rot is most prevalent upon marshy, low-lying, or stiff soils, and we frequently have experienced an outbreak when the flock has been removed from the uplands and brought into the richer and more unctuous portions of the farm. Lambs contract the disease at a very tender age, and are evidently much distressed with it. When this is the case advantage should be taken of the first dry day to "go round" them carefully, paring their little feet, and applying the remedy already mentioned.

FLIES.

Flies of various species are a great cause of annoyance to sheep and to shepherds. It is scarcely too much to say that a principal work of the shepherd during summer is to guard his flock against the attacks of flies. Of these I shall speak of three: (1) common flies; (2) the sheep bot fly; (3) the sheep tick fly.

COMMON FLIES.

It may not be entirely satisfactory in these days of science to speak of common flies without, at the same time, strictly defining what flies we mean. There is, however, a "very large division," as Miss Ormerod tells us, of the *Diptera*, or two-winged flies, all the species of which are more or less like a common house-fly. That the fly which torments sheep is precisely the same as the fly which crawls up our window panes, or torments our cooks in our kitchens, we may be sure is not the case. There is no doubt that insects which take up their abode in houses become in a measure domesticated, and hence the *musca domestica* of our houses varies considerably from the woodside hordes which appear to live for the sake of perse-cuting their fellow creatures, but doubtless, and in fact, as we know, have their uses in the economy of Nature. It is probable that if the swarm which has settled upon the head of a devoted sheep were examined by an entomologist he would easily iden-tify several species, all, however, very much alike. They are of the size and sombre hues of the house-fly, they are thick in the body, short in the proboscis, active on the wing, and we all know them as "flies." Buzzing around the tails, settling on the dung, and busily looking round for a place "where she may (literally) lay her young," are various forms of those larger and even more offensive creatures, the meat flies or blue-bottles, or bronze-coloured sisters, cousins and aunts of the same. These creatures, which are the horror of the pedestrian, the horseman and the shepherd, need no very special descrip-tion. They are associated with bright sunshine and intense heat, with country lanes flanked by woods, with pastures especially bordering on plantations, but less with open downs.

These creatures affect sheep at two vulnerable points—the head and the tail—and a few remarks upon them will be useful to breeders. So far as the head is concerned, we know from experience that the victims are principally the more or less bare-headed races of sheep. The Cheviot upon Lammermuir, or its native hills, the Border-Leicester, the Leicester, and the various Leicester-Cheviot crosses occur to us as particularly susceptible to this form of annoyance. The Down breeds are less affected, and, speaking from some experi-

14

ence, we should say enjoy a certain immunity from sore heads. Shropshire sheep, with their woolly helmets, Oxfords and Cotswolds with their characteristic fore-locks, Hampshires, and, to a less degree, Southdowns, are not plagued with flies about the head. It is at the roots of the small rudiments of the horn that flies first find out a weak spot. There they are no doubt first attracted by dirt and some form of decomposition which it appears Nature designs these creatures to feed upon and destroy. Any animal odour attracts them, and they become the cause of irritation, and this again induces a serious exudation which still further encourages attack. The flies now settle in swarms upon the devoted head. They trample with their tiny feet and perforate the skin for juices, until the effect is a broad scald extending over the crown. The poor animal in vain attempts to shake off his persecutors. He runs, he suddenly stops, he wags his tail, but not in pleasure; he stamps his feet, he throws himself down, and by shaking his head and ears, and taking some advantage of the bents which cover his pasture, attempts an ineffective resistance to his enemies. No sheep can fatten or thrive in this condition. His only relief is sunset or night; but in the early morning he is again subjected to slow torture. Such is the ailment—next for the treatment.

One of the essentials of treatment where large flocks are concerned is simplicity. Unless a remedy can be suggested which may be applied by the shepherd rapidly, and in accordance with his habits and notions, it is not likely to be accepted. It must be something which can be daubed on with a brush, sprinkled from a can, or readily applied by the hand. It must be thorough and immediate; and fortunately we can prescribe remedies of this nature. The simplest remedy appears to have been communicated to the public in Hogg's " Shepherd's Guide." " I happened," says the Ettrick Shepherd, " to be assisting at the sorting of sheep of the Cheviot breed, when sundry of their heads were broken with flies. The shepherds brought them out of the fold with the intention of smearing the sore parts with tar. I advised them to anoint them with coarse whale oil, such as they mix with the tar, having several times seen sores softened and healed by it. Some of it being near at hand they consented. The flies were at this time settled upon the fold in such numbers that when we went among the sheep we could with difficulty see each other, but those anointed with the oil were turned in among the rest, and to our utter astonishment, in less than a minute not a fly was to be seen." This remarkable experience certainly is announced as a discovery, for it surprised both the Ettrick Shepherd and his comrades. Whether it was entirely a new idea at that time it is impossible to say; but certain it is that whale or train oil is now looked upon as a sure safeguard against this form of fly attack. Experience has proved that a little sulphur mixed with the oil is beneficial, so that sulphur and whale oil, applied with a brush, will keep off the flies and prove a great advantage to the sheep. With such a simple remedy at hand, no one should be troubled with sore heads; and our reason for giving it is that, in spite of its simplicity, it is not universally practised or known.

The Sheep Cap.

The worst feature of the sheep cap is that when it becomes dirty or moist from perspiration the flies settle upon it and produce a sore under the cap. They thrust their proboscis through the calico, and suck up the animal juices, thereby producing as much irritation as before. It is therefore necessary to place a pad under the cap; but we prefer to dispense with this means of protection and to employ the simple plan first given. The sheep cap is made from a pattern which corresponds to the general shape of the sheep's head from between his ears to about half-way down his nose. There are two tape loops, one for each ear, and at the nose end of the cap there are two long and free tapes. On putting on the cap, it is placed over the head, the loops are placed over the ears, the two free tapes are passed on either side around the nose, between the eyes and nostrils, crossed under the jaws, passed through the two ear loops, and tied under the neck, just below the ears. Thus the ear loops are strained downwards on the cheeks, and the cap is held secure.

MAGGOTS.

That sheep may be literally eaten up by maggots is a known fact. The disease—for so it may be termed, being as much a disease as any other external or internal parasitic distemper—is serious. It varies in intensity from a mere annoyance to a fatal disorder. The flies which breed maggots are well known to be *viviparous*, or to bring forth their young alive. The maggots are thus deposited among dung, moisture, or dirt adhering to the coat, and they soon make their way through the wool to the skin. There they produce irritation, which is invariably followed with redness, soreness, and exudation of serous fluid. Thus a scald is established and the maggots, by incessant effort, succeed in burrowing into the skin itself. The sheep exhibits symptoms not unlike those we have described above, when he is attacked by flies about the head. He often assumes an attitude as though listening, his head is lowered, his ears pricked forward, and his feet are placed near together. Stamping, running, couching, and general unrest are the indications which attract attention. Sheep affected with maggots often lie apart from their fellows, and close to the ground. They from time to time jump up, run, and throw themselves down again.

The mineral known as mercury stone quickly kills maggots. Shepherds should always have a piece of this substance. It is as convenient to carry as a piece of slate pencil, which it somewhat resembles, and it may be purchased of any chemist. To part the wool and thoroughly rub the parts affected with mercury stone speedily kills the maggots, which ought, however, also to be brushed out by the hand until they are all got rid of. Spirits of tar may be used freely with the same good effect. Fluid mixtures compounded for the purpose may also be purchased of any druggist in a sheep-farming district. The colouring matter which is applied to rams or show sheep is often ostensibly applied as "fly powder." The object, no doubt, is twofold, as the powder is coloured according to taste, and is employed as a means of decorating an animal intended to appear before the public. It, however, possesses an actual use, as it keeps away the flies from animals which it is particularly desirable should not be disfigured in any way.

When maggots are neglected the sheep pulls out his wool with his teeth, and, later, the skin becomes hard and dry, and the wool comes off very much in the same manner as if the skin had been scalded or burnt. The use of fly powder is relied upon to prevent these injurious effects, which would ruin any sheep intended for exhibition, or for appearing in a sale ring as a ram.

The Sheep Bot Fly (Œstrus Ovis).

This fly belongs to the same order as the Horse bot fly (*Gastrophilus equi*). It is rather larger than the common house fly, and of an ashy colour, spotted with black (Ormerod). It appears in May, June, and July, and becomes an intolerable nuisance to sheep. Youatt tells us " that if only one appears the whole flock is in the greatest agitation." The peculiar instinct of the *Œstrus ovis* is to deposit its eggs on the inner margin of the nostril, when they soon hatch, and at once proceed to crawl up the nose, until they reach the recesses of the frontal sinuses, where they hang on with their tentaculæ until the following May.

During their passage up the nostril, and until they settle themselves in their extraordinary habitation, they give great annoyance to their unfortunate hosts, but after fixing themselves they do not appear to cause inconvenience or mischief. Again, when they are ready to descend by the nostril, in order to change into the pupal form, they give rise to much pain and violent sneezing and signs of discomfort. It is not often that more than three bots are found lodged in one sheep's head, although a dozen have been recorded. The best way of seeing the bot is to examine sheep's heads in April or early May by splitting them and searching among the convolutions of the frontal sinuses. The bot often has reached the cancellous or spongy recesses which are found at the base of the horn—these, as is well known, being in communication with the air passages and higher convolutions of the nasal cavity. When the worms are caught in the act of expulsion from the nose, or are taken in that advanced state from the cavities of a newly-killed sheep, they are very restless, and are continually marching, or rather dragging, themselves along. When placed upon the hand they find their way to the divisions between the fingers, and, using the points of their crochets, they endeavour to force them apart. They soon get to the bottom of the loose earth or powder in the insect box; and if they are placed on the ground they very speedily bury themselves in it, and are lost.

The sheep bot fly is by no means easy to encounter with preventive measures, but it is probable that the same odours which drive away other flies might be equally distasteful to the *Œstrus ovis.*

Gid—Sturdy—Turnsick.

One of the most strange and not the least fatal malady which affects our lamb flocks is that peculiar parasitic affection known to shepherds and flock-masters as "gid," "sturdy," or "turnsick." These names, like many others employed in the nomenclature of disease, do not in any way indicate the nature of the ailment, but are merely expressions denoting a leading symptom in the disorder. Animals suffering from the disease are invariably giddy and turn round to the right or to the left, or obstinately stand as if fixed to the spot and refuse to move on ; hence have arisen the terms by which the malady is commonly known. Gid is essentially a form of animal parasitism, and owes its origin to the presence of one or more cystic or bladder-like parasites in the substance of the brain, and to this fact the giddiness and disordered movements of the infected lambs are due.

How these cestoid worms, as they are termed, obtain an entrance to the body can only be shown by relating a short biological story. The gid parasite consists of an hydatid, or sack, filled with a watery fluid, and is sometimes spoken of as a bladder-worm. It varies in size from a hempseed when young to that of a cricket ball when matured. These bladder-worms are derived from a tapeworm (*tænia cænurus*) which infests the dog, and are in reality the young or larvæ of that creature. Just as the maggots which we find in flesh represent the young of the blow-fly, so do these hydatids in the brains of lambs represent the infant stage in the development of the tapeworm. Each hydatid, however, represents not one but a brood of many larval tapeworms. How many may be known by counting the little white spots which are seen scattered over the wall of the bladder, and which are in fact the heads from which future tapeworms are to spring should they find their way into the bowels of the dog. Without the dog and other members of the species to which he belongs there could be no gid parasites, and consequently no giddy sheep. The lamb acts towards the bladder-worm as a host, *i.e.*, it affords it shelter and nourishment in its youth until fitted for its new habitation in the dog, where it proceeds to throw out segments and complete its development into tapeworms. Without sheep and others of the species to nurse the bladder-worm, there could be no tapeworm. Neither the hydatid nor the tapeworm can be regarded as welcome guests. The former by its growth and expansion breaks down the brain and paralyses its host, while the latter reduces the dog it infests to starvation by disordering the digestive canal, and not unfrequently proves suddenly fatal by exciting fits of epilepsy.

How hydatids gain access to the body of the sheep has been clearly established both by experiment and observation. Hounds, shepherd and other dogs roaming over pastures and root fields commonly leave behind them matured segments of tapeworms in their excrement. Each of these little flat white fragments, with which all dog owners are familiar, is loaded with eggs, which soon escape on to the ground, and having become free are swallowed with the food of the sheep, or failing to be thus safeguarded perish and decompose. Safely landed in the intestinal canal the period of incubation commences, and soon each egg discharges a minute embryo, which, by some means as yet unexplained, finds its way to the brain, there to develop into a bladder-worm or gid parasite. In order that this creature may become a tapeworm, it must, as we have remarked, pass back into the dog. Many, no doubt, fail to return to their canine host, as many tapeworm eggs fail to reach the sheep. Some perish in the brain of the lamb, and degenerate into a pasty mass, in which case the host triumphs over the guest, but more frequently the sheep succumbs to the destructive influence of the parasite. It is then that further development of the latter is rendered possible, for the dead sheep may become food for the dog, and should the head with its living contents be eaten, then development of the hydatid brood into tapeworms is assured, and the life cycle of the parasite rendered complete. If, however, the head of the infested sheep be not consumed by dog or fox, or some other suitable creature, then the bladder-worm perishes, and decomposes with the brain which contains it. This is, no doubt, the lot of many, or gid would over-run our flocks to a much greater extent than now.

There are few shepherds of experience who are not familiar with the eccentricities of sturdy sheep. The onset of the disease varies in different cases, and the symptoms are by no means alike in all. In some the patient is dull, and hangs back from the troughs. The eyes are glassy, and the face wears a vacant stare. It may obstinately stand, and require to be pushed along. The approach of the dog is disregarded, or but little noticed, and such an animal is often said by shepherds to be "stunned i' th' head." In other instances there is added to these symptoms various forms of disturbance of locomotion. Some, while being driven along, turn more or less towards the right or the left. Many describe a circle in the one direction or the other. Others roll about from side to side, and are, as the name of the disease suggests, giddy.

In some instances the animal moves forward in leaps and bounds not unlike the skipping action of a deer. Others, again, obstinately back, and are altogether incapable of moving in any other direction. Besides disturbances of locomotion, the muscles of different parts of the body are sometimes seen to twitch, or the entire body may become convulsed.

If the head of an animal so affected be carefully manipulated, a slight bulging of the bones of the skull may sometimes be recognised over the place where the parasite is lodged, and by a little pressure with the finger the bones at this part will be found to yield, owing to their having become thinned by the outward pressure of the parasite within. The hydatid, and consequently any bulging which may exist, will invariably be found on that side of the skull towards which the animal moves, whether to the right or to the left. When the movements are such as we have described as taking place in a forward direction, the hydatid must be sought for somewhere in the middle of the head.

The treatment of gid is rarely attended with complete success. It necessitates, of course, the removal of the parasite from the brain, which can only be effected by laying open the cavity of the cranium. The practice of puncturing the bladder with a pointed instrument and drawing out the fluid with a syringe is relied on by some, but it is seldom that more than temporary relief is afforded by the operation. The mischief done to the brain before the presence and the seat of the parasite can be made out, and the frequent necessity for a second operation, are serious drawbacks to success. We have tried both modes of operating, but the results do not encourage us to place much reliance on surgery as a curative measure.

Prevention in this, as in other kindred affections, is the lesson which most commends itself to flock-masters.

We have already pointed out the relationship subsisting between the gid parasite and the tapeworm (*tænia cænurus*) of the dog, and it will be evident that to guard against the one entails the exclusion of the other from the farm. In the spring of the year in particular, dogs on the farm should be carefully watched, and on the appearance of tapeworm segments in the excrement they should be placed in confine-ment at once and deprived of their liberty until the parasites have been altogether got rid of. To accomplish this a dose of areca nut, or some other suitable vermifuge, may be given while fasting, to be followed by a dose of castor oil after the lapse of an hour. Should one dose not have the desired effect, a second must be administered, and, if necessary, a third. When expelled, every particle of the parasite should be burned, or otherwise effectually destroyed.

It may be necessary to repeat that only one particular tape-worm (*tænia cænurus*) is capable of propagating the gid hydatid, but as the characters by which it is distinguished are unknown to flock-masters, it is in the interest of the dog, no less than the flock, that tapeworms of whatever kind should all be destroyed.

To render the system of prevention complete necessitates not only the destruction of the tapeworm, but also the gid organism. Dogs must not be allowed to eat the brains of infested sheep. Whether the latter be slaughtered, or die from the effects of the disease, the head with its contents must be well boiled and buried.

With all these precautions, however, gid will continue to prevail, especially in hunting districts where hounds cross the land, and as they, no doubt, do leave behind fragments of the offending tapeworm, masters of hounds might do much in the interest of sheep farmers by urging upon the attention of their kennel huntsman the desirability of keeping the pack as free from tapeworm as possible.

Pastures near to towns crossed by public footpaths are specially prone to become infected with the eggs of tape-worms from dogs which pass over them. Such enclosures should not, if possible to avoid it, be stocked with lambs, but reserved for older sheep.

Joint-ill (*Pyæmic arthritis*).

The early life of the lamb is threatened with many and various diseases which too often visit the fold early in the year, and thin down to ruinous proportions the future of the flock.

Not the least important of these terrible visitations is that acute and destructive affection to which custom has assigned the term "joint-ill," and which is generally regarded by shepherds and flockmasters as having some obscure connection with the "weather." It should, however, be pointed out that while the affection so termed usually develops changes of a marked character in the articulations of the limbs, such alterations are by no means invariably manifested in the course of the disorder. It is frequently the case that this so-called "joint-ill" assails and destroys its victims without exciting any obvious changes whatever in those parts from which it has derived its name. While, there.ore, joint disease may be admitted to form a leading symptom of the malady, it must be understood that the disease itself is one of a general systemic nature, of which the joint affection is but a local manifestation. In whatever form it presents itself it frequently kills, or so far depletes and wrecks the system as to render the survivors from the affection most undesirable property.

The circumstances and conditions under which it arises were, until recently, most imperfectly known, and hence it was that pathologists, both in this and other countries, were widely divided in their opinions as to the nature and cause of the affection. In this connection various views have been entertained by different observers. Rheumatism was for a long time accepted as the inducing factor until that disease became better understood, and its many points of departure from the malady in question were fully recognised.

The existence of abscesses in and about the joints, as well as the liver, lungs, and other organs of the body, led to the suspicion of a scrofulous nature, but as more light fell upon the latter disease, its relations with joint-ill disappeared, and are no longer recognised.

So far as the conditions of its origin are concerned, there is every reason for believing that it does not owe its existence to any one particular cause, but rather to the operation of several co-ordinating circumstances connected with the feeding and management of the ewe flock as well as the offspring.

It may be noted that the disease is confined to no particular breed. The offspring of old and young stock are alike affected, but it must be admitted that the lambs of ewe tegs suffer much more frequently and severely than those of older sheep.

In both old and young the liability to attack is greater in twins than in singles. Of the influence of locality we have no sufficient data upon which to form a judgment, but of meteorological conditions it can be said that cold easterly and north-easterly winds, especially if attended with rain or snow, aggravate the malady when once it is established, but it does not appear to have any notable influence in its production. We have known it to prevail under the most favourable conditions of weather. From twenty-four hours to three weeks or a month old is the time when the disease usually appears, and its duration extends over a like period. Isolated examples of the affection seldom occur. Large numbers of the flock invariably suffer, and of those stricken 40 to 80 per cent. succumb.

Vague references have now and again been made to hereditary influence, but no data are forthcoming to establish it as a factor in the induction of the disorder. Nor is there any reason for regarding it, as some have done, as a contagious affection.

Gamgee found that flocks affected with this disease " had been kept in a confined space in winter, had been fed well, and not allowed to move about sufficiently." In our own experience this has not been the case. We have seen it in open folding with plenty of room to " fall back," but where the food was confined exclusively to roots until a short period before parturition. It has mostly been the case that the roots have been of indifferent quality and much diseased, and the ewes in low condition when brought into the lambing pens.

Without ascribing the disease altogether to a state of debility, such as an exclusive diet of indifferent roots may induce, we are strongly impressed with the importance of this condition as a predisposing influence, leading, as it must, to a general debility and constitutional impairment of the fœtus.

For a better understanding of the exciting cause, we turn to the state of the navel, where we usually find more or less swelling, or some imperfect closure of the orifice, or an abscess, or a soft putrefying blood clot occupying the mouth of the vein which in the fœtus passes from the umbilical orifice to the liver, and there is every reason for the belief that the incomplete closure of this fœtal orifice, arising out of an indifferently nourished organism, serves as a channel through which matter of a putrefactive character gains access to the circulation, and, settling down in the joints and other organs of the body, excites in them inflammation and abscess. Hence the disease has come to be regarded as a state of pyæmia or poisoning of the blood through the imperfectly closed and diseased navel opening. This view of the pathology of the affection seems to explain much, both in the symptoms exhibited during life and the changes observed after death, which could not otherwise be accounted for.

It is hardly necessary to point out that where this tendency to imperfect healing and suppuration of the navel exists, dirty lambing pens and dirty shepherds supply all that is needful to contaminate the wound, and start a centre from which the blood stream may be fatally polluted.

The symptoms of the disease are very striking and characteristic. The affected lambs are dull and lowering, and lie about with little or no disposition to feed. Soon stiffness appears in one or more of the limbs. This is quickly followed by severe lameness and swelling of the joints, mostly affecting the knees and hocks first, then spreading to the others. The enlarged joints are at first firm, hot and painful to the touch, but later on they present small soft fluctuating points, which sometimes break and discharge a thick ropy pus. Not infrequently abscesses form under the throat, on the arms and thighs, and in various other parts of the body.

Fever, indicated by thirst, shivering, and a high temperature, is always present to a greater or less degree. In some instances, as we have already pointed out, the joints escape altogether. In these cases the disorder shows itself by great and sudden prostration, inability to stand, high fever, hurried, panting breathing, and, later, diarrhœa and fœtid breath, to which is sometimes added a tympanitic state of the belly, and a yellow condition of the membrane of the eyes.

With regard to treatment, it may be stated at once that very little can be done to eradicate the disease in animals so young when once it is established. The attention of the flockmaster must be concentrated in the direction of arresting its spread, and this must be done by giving prompt and special attention to the ewes and the condition of the lambing pens. In connection with the former they should be placed on a liberal ration of good nutritious food, and if large quantities of turnips are being allowed the quantity must be reduced, while at the same time such as are diseased should be as far as possible avoided. As a matter of course, the healthy will be separated from the sick. If ewes are still lambing down it will be desirable to thoroughly cleanse and disinfect the pens, or even to remove them to another part of the field or farm-stead. As to medicine, salt and sulphate of iron given to the ewes morning and evening in their manger food will aid in imparting tone to the system and improving the general health.

All lambs born after the appearance of the disease in the flock should have their navels carefully examined and dressed for the first two or three days, or longer if necessary, with a solution of carbolic acid, and until turned loose should be folded on clean dry litter.

Fouling of the pens should be carefully guarded against, and cleanliness on the part of the shepherd ought to be strictly enforced.

The affected lambs will require to be confined in pens with their dams, but it is not advisable to resort to any medicinal treatment beyond a small dose of castor oil where constipation of bowels exists, or a little carminative astringent in the case of diarrhœa. The latter may be provided by mixing together a small quantity of prepared chalk and powdered nutmeg, and adding to it a small teaspoonful of brandy; the whole to be given in milk. Any good that may be done must be effected chiefly through the milk of the dam, whose general health it should be the aim and object to improve by a generous dietary and ferruginous tonics.

Aphthæ—Thrush.

No less troublesome and fatal than the disease to which we have referred is that disorder of the mouth, commonly known as "thrush" or "canker," which so frequently prevails in our lamb flocks during the spring months.

Aphthæ is an eruptive inflammation affecting the lining membrane of the mouth, and frequently extending also to the skin covering the lips and face. It is similar in character to a disease of the human family, and with which most parents are more or less familiar.

Young lambs from two to six weeks old are mostly its victims, but older sheep are sometimes seriously affected by it. It rarely, however, prevails in stock over two years old. As a rule the fatality arising from this disease is very considerable in sucking lambs, and we have known it on more than one occasion to sweep off a large proportion of a promising flock.

The malady is not limited to any particular localities, but prevails from time to time wherever sheep husbandry is carried on. Our experience, however, seems to indicate its larger presence in strictly turnip countries.

Seasonal conditions appear to exercise a marked influence over its development. A mild winter and temperate open spring, when turnips shoot into luxurious growth at the period of lambing, and meadows put up rank innutritious grasses, have certainly coincided with the most widespread outbreaks of the affection we have experienced.

As to the immediate cause of the disease but little is really known. It has been stated to be a strictly dietetic affection. At the time of its occurrence lambs are living exclusively on the mother's milk, and the malady is generally regarded as arising out of some deleterious property which this secretion has had imparted to it by the bodily condition of the dam. It is quite true that ewes whose offspring suffer from this disorder are generally in low condition and wanting in vigour and bloom, and it is not infrequently the case that they afford indications of impairment of the digestive organs by a loose and unduly active state of the bowels. We have on several occasions found it to arise where ewes, badly wintered, have been allowed to run over turnips about the month of March, and fill themselves with rapidly-grown succulent sprouts.

The unrestrained use of such a diet is well calculated to disorder the digestive system, and lead to impairment of the lacteal secretion. Sudden changes from roots to the various forage plants provided for spring feed for the ewe flock are possibly also in some way connected with outbreaks of this affection; at least, this has appeared to be the case in such inquiries as have from time to time engaged our attention. The fact cannot, however, be lost sight of that the presence of this disorder in lambs rarely or ever exists alone, but is mostly attended with a form of inflammation of the udder of the dam, which, on account of its destructive effects on the gland, has been designated malignant or gangrenous mammitis.

This disorder prevailed to some extent in the ewe flocks of Lincolnshire during 1889, and proved very fatal to both lambs and ewes.

The fact that it is communicable from the diseased mouth of the lamb to the teats and udder of the ewe, clearly establishes its contagious nature, and M. Nocard, who has studied the micro-pathology of the affection in the latter, avers that it is due to a minute organism or micrococcus, which he has found not only in the milk but also in the watery effusions present in the abdominal cavity after death.

That this organism is the specific cause of the disease is shown by its capability to induce the affection in previously healthy stock, whether inoculated with the milk itself or after being separated from it by the usual methods of cultivation. Professor Brown, of the Agricultural Department, when studying the disorder as it occurred in the Lincolnshire outbreak, has recognised in the diseased tissues of the lamb, a microbe in every respect similar to that discovered by M. Nocard in the ewe.

It is therefore evident that the disease as it occurs in parent and offspring is the same, and apparently the work of the same micro-organism.

In our experience, the lambs have invariably given evidence of the disease first, and afterwards the ewes have appeared to become inoculated from the mouths of their young; but as to the source from whence the lambs receive the infection, that is a question still waiting solution.

Is the *materies morbi* primarily contained in and given out by the milk of the dam, or is it only after the teats become inoculated by the lamb that it gains entrance to the gland?

15

If the milk is not contaminated in the first instance, then the source of the infection must be sought for elsewhere. One possible alternative appears to be that it may be sucked from the surface of the solid teat and ingested with the milk, but from what source the teats become soiled there is no evidence to show.

Referring to the symptoms of the disease, the first indications of illness are shown by the lambs hanging back from the ewes. When at the teat, they refuse to suck or grasp it with the mouth, and, after a few attempts, liberate it again. In the act of sucking, the mouth becomes filled with foam, which hangs about the lips and udder of the dam. The youngsters cease to play, and lay about the pastures. Weakness and wasting quickly become apparent. The head droops, the ears hang pendulous, and the movements are feeble and unsteady. At this time the outer surface of the mouth becomes covered with small red pimples, varying in size from a hempseed to a pea. Soon vesicles or small blebs appear on their summits, and later these develop into sores, and become covered with scabs.

A similar eruption is found to exist in the mouth, especially on the tongue, gums, and lips. When severely affected, the gums ulcerate, or throw out proud flesh, and the teeth become loose, and fall from their sockets. In such cases the lower jaw bone seldom escapes disease and more or less disorganisation. This complication altogether disables the animal from feeding and adds considerably to the mortality.

Inflammation of the lungs, followed by abscesses, not uncommonly arises in the course of the affection, when great prostration, inability to move, quick panting breathing, frequent coughing, and other serious symptoms are presented.

As previously noticed the udder of the dam almost invariably throws out an eruption similar to that in the mouth of the lamb. Sores cover the teats, and these again are crusted over with thick black scabs. Sucking on the part of the lamb is now out of the question, and the milk being retained serves, with an extension of the disease along the milk duct, to induce garget or inflammation of the gland, which not infrequently results in the sloughing of one half or the entire organ.

Nothing in the whole range of veterinary practice is so difficult as to prescribe for the successful treatment of cases of this description. To prevent the spread of the disorder is of the first importance.

In this connection isolation of the sick should be promptly carried out, and the shepherd should have no communication with them until they are again convalescent.

An entire change of food as well as a change of ground is most important as regards the healthy part of the flock. The diet must be nutritious, plentiful, and stimulating, and, where roots are to form a part of it, they should be sound and wholesome and sparingly given.

With reference to the sick, the affected lambs may continue to suck the ewes so long as their udders remain free from disease, but so soon as the eruption presents itself on the latter they should be taken away and fed with the bottle, while at the same time the ewes are carefully hand-milked until the gland is restored to a normal condition. This we are aware entails immense trouble, but it is the only way by which losses amongst ewes can be successfully guarded against.

A saline aperient, such as a little Epsom salts combined with some aromatic as aniseed or coriander seed, may be given to the sick ewes at the outset, and this should be followed by the administration of a little powdered columba root and salt in the manger food twice daily.

As a dressing for the udder a 5 per cent. solution of carbolic acid applied morning and evening will assist in healing the wounds and dispersing the eruption, or, what is equally efficacious, a weak solution of alum.

The mouth of the lamb both inside and out should be washed two or three times a day with a solution of chlorate of potash of the strength of ten grains to the ounce, using just sufficient to moisten the surface without allowing any to be swallowed.

Where proud flesh appears on the lips or gums, it should be promptly touched with lunar caustic, and any teeth which may become loose must be removed.

LIVER ROT, COATHE OR BANE.

No disease is more dreaded by the flockmaster than rot. Although the true nature of this fearful malady has only been understood during recent years, its fatal character has been long known under various names. We must necessarily write with brevity upon this subject, although the material at our disposal might be extended over many pages. The proper understanding of the disease is in a great measure due to the investigations of Mr. A. P. Thomas, which are recorded in vols. xvii., xviii. and xix. of the Royal Agricultural Society's Journal (second series). It is not necessary in this connection to do more than give a general and popular account of this extraordinary and fatal disease which claims, it is stated, 1,000,000 victims a year in Great Britain. We can never forget the devastation of our flocks after the wet summer of 1879, which was followed by what was at the time called a sheep famine in 1882 and 1883. This is the fellest of all ovine diseases, and is due to the presence of parasites in the liver of the sheep. The disease is produced by an invasion of what is popularly known as FLUKE, and scientifically as *Fasciola hepatica*, or *Distomum hepaticum*, a member of the Trematode section of the Annuloida (Steel). It is a flat worm, having a general resemblance to a flounder, by which name it is also known.

One of the most interesting and fascinating pages of modern natural history concerns the life history of the entozoa or internal parasites of mammalia, of which the fluke is a striking example. Like so many entozoa, recent researches have shown that their life history depends upon their entrance into more than one " host," or receiving animal, at different stages of their existence. Of this fact there are many examples, such as the tapeworm and the bladder worms, which cause gid or sturdy in sheep.

The liver fluke follows this dual existence, and at different stages is found inhabiting the body of a mollusc, shown by Thomas to be the *Lymnæus truncatulatus*, a freshwater snail with a brown spiral shell, which is very common and widely distributed, but has no popular name. The late John Henry Steel informs us in his excellent treatise on "Diseases of the Sheep" that its shell is sometimes under a quarter of an inch, and never over half an inch, in length, and Thomas found that these snails suffer much from the invasion of the fluke in a certain stage of their existence, and that this fluke disease is as fatal to them as to the sheep. The invasion takes place, so far as the snails are concerned, in April and May, but more generally in June and July. " A severe winter kills off ' rotten snails,' but a mild winter may simply render them torpid, and thus unusually early cases of rot in sheep may occur in the following May. In the tissues of this snail it develops into the sporo-cyst, redia, or ' nurse,' a kind of sac in which develop numerous larvæ with tails (cercariæ). These escape through a special vent in the redia, and for a time lead independent free lives in water. They then come to rest, cast their tails, and develop an enveloping cyst of a snow-white colour which adheres to the stalks or leaves of grasses and water plants. These may remain a few weeks, but if they undergo no further change of conditions the embryo within perishes. In this form on grass or in water they pass into the alimentary canal of the host or ultimate bearer (sheep, horse or other mammal), and the pupa-cyst is digested by the action of the gastric juice, and from the stomach of the host through the duodenum and bile duct the fluke passes gradually up to the liver." Whether this is an absolutely correct account of the development of the fluke may be doubted, as Cobbold remarks that " there is no recognisable limit to the variety or to the extent of larval fluke development." To speak generally, there appears to be no doubt that by some such process as that described the fluke passes from the snail into the digestive system of the sheep, and from thence into the bile ducts, where it arrives at the stage of sexual development or maturity.

Sheep receive flukes into their system in July, August and September, or the later summer months, and the virulence of the disease depends upon the number of flukes which find their way into the digestive system of the sheep. If few flukes are present no injury may follow, but a number sufficient to crowd the bile ducts proves fatal. The subject is full of difficulty, for the degree of mischief depends upon the numbers of the invading parasites. The fluke does not propagate within the sheep, although it lays innumerable eggs, which, with the excrement of the flukes, fill the ducts and cause fatal derangement of the system. It has been ascertained that flukes do not multiply within the sheep, and that the severity of the attack depends upon the number of sporocysts taken into the body of the sheep. These, when they have attained maturity, lay eggs which are expelled by the flukes in all parts of the bile ducts, in the intestinal canal, and on the soil after escape of the flukes from the body, but wherever they are deposited, their ultimate destination is the soil, to which they are conveyed, either with the fæces as free ova, or in the mature fluke as expelled from the sheep. The infested sheep becomes a bearer and distributor of flukes as he moves from pasture to pasture, expelling ova. As one sheep may contain a thousand flukes, and each fluke forty thousand eggs, it is manifest that one bearer may contaminate a pasture, though that bearer be in apparent health.

The cause of fluke disease in sheep is, then, an accumulation of encysted larvæ in the body of the sheep, which there develop into mature flounders or flukes, capable of laying eggs within the sheep, or without, if expelled from the body. An infested sheep, therefore, is a source of contamination by stocking the pasture with eggs and ova. The flukes are probably for the most part expelled from the host (the sheep) when mature, and when depositing their ova.

Rain washes the ova into drains, ponds and other accumulations of water. The egg contents develop into ciliated embryos which swim about freely in the water. It has the figure of an inverted cone, with a proboscis-like papilla at the centre of its broad and flat anterior portion. After a few days of free, rapidly-moving existence the young creature loses its cilia and becomes a creeping larva, and thus finds its way into the mollusc which is the first intermediary bearer of the fluke. Thus the life circle is completed, and after the sporo-cyst is formed, the new generation again completes the vicious circle of its existence, provided it finds a suitable host in a sheep. If not, it dies. (Steel.)

The four stages of fluke disease have been enumerated as follows by Gerlach:—

(1) Traumatic hepatitis, which occurs between June and autumn, is seldom diagnosed but sometimes seen after death from apoplexy. The liver is in a state of acute inflammation.

(2) Dropsy. The liver pale and firm. Occurs from September to November or December, *i.e.*, six to twelve weeks after invasion.

(3) Emaciation, the stage of greatest mortality. It occurs in January and February, and the liver is atrophied.

(4) Migration of the flukes occurs in May and June, but Thomas has evidence of flukes remaining in the sheep longer than a year.[*]

When sheep are first invaded by the parasitic sporo-cysts there is a period during which the host appears to be benefited and to accumulate fat. It is for this reason that butchers and graziers sometimes put sheep on to unsound land shortly before slaughtering. It is not necessary to consider this as a stage of the disease, as the action of the flukes may be stimulating to the system, and a small number of flukes have even been held to be beneficial. Rot appears to be due to an excessive number of flukes rather than to the presence of a few, and the complete breaking up of the constitution of the host is due to numbers. In this connection it is well to remind the reader that the flukes are incapable of multiplying

---

[*] Steel, " On Diseases of the Sheep."

in the sheep, but are all and each imported from the infested grass. The close biting of sheep is supposed to be one cause of their receiving such vast numbers of sporo-cysts, and hence long grass is not so fatal as closely grazed infested meadows.

TREATMENT OF ROT.

The treatment is rather preventive than curative. Frost and drought are natural safeguards, and thorough drainage of land is a great assistance in destroying the conditions favourable to the development of ova. Wet spots should be fenced off, and the grazing of sheep be carried out upon " sound " land. Neither must it be forgotten that infested sheep spread the disease, and ought, therefore, to be carefully separated or destroyed. The strength of sheep ought to be maintained by plenty of dry and concentrated foods, and tonics such as salt and sulphate of iron may be given with advantage.

To cure a rotten sheep is more difficult, and is looked upon by many as hopeless. If, however, the animal is enabled to outlive the invasion and exist until the expulsion of the flukes it may recover. Professor Simond's recipe will prove a good guide, and is as follows: Take of bruised oil cake and pea meal of each a bushel; of salt and aniseed each 4lbs.; of sulphate of iron 1lb. ; finely grind these substances, and thoroughly mix them and give to each sheep from half to one pint daily. Salt is an excellent tonic, and assists in the secretion of healthy bile, and it has been long known that sheep grazed upon salt marshes are never affected with rot. Change of diet is also insisted upon, and animals affected may receive stimulants such as turpentine and sulphuric ether. Succulent herbage should be avoided. The line of treatment consists in fortifying the system in every way, by good hygiene, tonics, and stimulants, giving food of an easily digestible character, and such as can be digested with the minimum of bile (Steel). Treatment is often unsatisfactory, and the best is undoubtedly that which keeps a flock as far as possible out of harm's way, by avoiding the conditions which are likely to develop this disorder.

INDEX.

ONLY ONE ADDRESS.

Day & Sons, Crewe.

Established 1840.

HORSE. CATTLE. SHEEP & DOG MEDICINES.

Largest Veterinary Providers in the World.

"Breeders and Owners of Stock can rely on the Preparations supplied by Messrs. DAY & SONS, of Crewe."—LIVE STOCK JOURNAL.

DAYS' "BLACK DRINK"

(Often called the "Magic Drink") cures like a charm Colic or Gripes and Chills in Horses and Cattle; instantly relieves Hoven or Blown Cattle and Sheep; stops Scour in, and is the best general Stimulant and Tonic for Calves and Lambs.

Matchless as a Restorative and Painkiller after Lambing and Calving; for Fatigue in Hunters and overworked Horses, and in all cases where nature flags.

Price 5/- per Quarter Dozen, post paid, or 19/- per Dozen Bottles in Boxes. Carriage paid.

DAYS RED DRINK,
Or Cow Drench.

For Costiveness, Loss of Cud, Carget, Colds, Fever, Hide-bound, &c. Prevents Milk Fever, and Cures Bad Cleansing. Prices—**1/-** per packet; for Ewes, **3/6** per dozen.

ARRANGED FOR
Disorders of
Horses, Cattle, and Sheep.

Prices:

£5, £2 4s. and £1 4s.

(with Guide, "Everyday Farriery").

—

Carriage Paid.

ARRANGED FOR
Disorders of
Horses and Colts.

Prices:

£5, £2 14s. and £1 4s.

(with Guide, "Everyday Farriery").

—

Carriage Paid.

Special Lambing and Calving Cases £1 1s., £2 2s., £3 3s. and upwards, carriage paid.

Note that we have only "One" Address, **DAY & SONS, CREWE.**

Advertisements.

Flockmasters secure better Breeding results and reduced mortality amongst lambs

BY USING

J. RANDS & JECKELL'S

SHEEP AND LAMB

SHELTERING CLOTHS.

SHEEP AND LAMB

SHELTERING CLOTHS.

'Indispensable to every Flockmaster.'

20 yards long, 3 feet deep, with Brass Eyelets and Cords for fixing to Hurdles.

6d., 9d. & **1s.** per yard.
Rot Proof **1/6** per yard.

THE "DUPLA"

SHELTERING CLOTHS.

"The Flockmaster's Friend."

20 yards long, 6 feet deep, with Brass Eyelets and Cords for fixing to hurdles.

1s., 1/6 & **2s.** per yard.
Rot Proof **3s.** per yard.

Carriage Paid on Orders above £2; 5 per cent. Discount for Cash.

WRITE IMMEDIATELY FOR NEW PAMPHLET ON

"SHELTER FOR SHEEP AND LAMBS."

Post Free on application to

J. RANDS & JECKELL, IPSWICH,

Sack, Rick-Cloth and Tent Manufacturers by Special Royal Warrant to H.R.H. The Prince of Wales.

LIVE STOCK JOURNAL.

Friday. (Illustrated.) Fourpence. Established 1874.

The only Paper in the United Kingdom wholly devoted to the Interests of Breeders and Owners of all varieties of Live Stock. Contains contributions from the highest authorities on all matters relating to the Breeding, Feeding, and Veterinary treatment of Domesticated Animals, and Illustrations of the more celebrated specimens.

Gives the fullest and earliest reports of Agricultural Shows, Stock Sales, Sheep Sales, and Lettings, whilst its Herd and Flock Notes and Notes from the Stables contain much valuable and interesting information. Prominence is given in the columns of the JOURNAL to correspondence on all questions of interest to Country Gentlemen, Breeders, and Exhibitors.

**Subscription:—3 Months, post free, 5s.; 12 Months, 19s. 6d.;
Foreign Subscription, £1 2s. per year.**

AGRICULTURAL GAZETTE.

A WEEKLY JOURNAL OF FARMING & MARKET GARDENING.

MONDAY. (ILLUSTRATED.) TWOPENCE.
Established 1844.

Has for many years stood at the head of the English Agricultural Press. Unequalled as a comprehensive practical paper. All branches of farming —crops, live stock and dairy—are fully discussed by leading practical authorities. Market intelligence and reviews of the grain and cattle trades are special features. Prompt replies given to questions in all departments of farming. Veterinary queries answered by a qualified practitioner. The Market Gardening section deals fully with the production and Marketing of vegetables and fruit. Special Articles on Cultivation, Manuring, New Varieties, &c., appear weekly.

Subscription:—3 Months, post free, 2s. 9d.; 12 Months, 10s. 10d.

BAILY'S MAGAZINE
OF
SPORTS & PASTIMES.

Racing, Hunting, Shooting, Yachting, Rowing, Fishing, Cricket, Football, &c.

This well-known monthly contains articles written by the best authorities on every phase of British Sport; and in addition to the usual Frontispiece —a Steel Plate Portrait of an eminent sportsman—other Illustrations of well chosen subjects and of the highest artistic merit are given.

Of all Booksellers and at all Bookstalls, 1s. Or by post direct from the Office, 14s. per year.

India Proofs of any of the Engraved Portraits of which some 500 have appeared, 2s. 6d. each.

VINTON & CO., LTD., 9, New Bridge Street, London, E.C.

LIVE STOCK HANDBOOKS.

No. 1.

SHEEP: Breeds and Management.

SECOND EDITION.

By JOHN WRIGHTSON, M.R.A.C., F.C.S., President of the College of Agriculture, Downton; Professor of Agriculture in the Royal College, London, &c. 236 pages, demy 8vo, cloth, gilt lettered, with 24 full-page Illustrations of the various Breeds. 3s. 6d.; post free, 3s. 10d.

No. 2.

LIGHT HORSES: Breeds and Management.

SECOND EDITION.

By W. C. A. BLEW, M.A.; WILLIAM SCARTH DIXON; Dr. GEORGE FLEMING, C.B., F.R.C.V.S.; VERO SHAW, B.A., &c. 226 pages, demy 8vo, cloth, gilt lettered. 28 full-page Wood Engravings of the various Breeds. 3s. 6d.; post free, 3s. 10d.

No. 3.

HEAVY HORSES: Breeds and Management.

SECOND EDITION.

By HERMAN BIDDELL; C. I. DOUGLAS; THOMAS DYKES; Dr. GEORGE FLEMING, C.B., F.R.C.V.S; ARCHIBALD MacNEILAGE; GILBERT MURRAY; and W. R. TROTTER. 224 pages, demy 8vo, cloth, gilt lettered, with 29 full-page Illustrations.

No. 4.

CATTLE: Breeds and Management.

By WILLIAM HOUSMAN and Professor J. WORTLEY AXE. 272 pages, with 34 full page Illustrations.

No. 5.

PIGS: Breeds and Management.

By SANDERS SPENCER and Professor T. WORTLEY AXE. 180 pages with 20 Illustrations.

3s. 6d. each, or post free 3s. 10d.; or the set of 5 volumes, if ordered together direct from the office, 17s. 6d., carriage free.

VINTON & CO., LTD., 9, New Bridge Street, London, E.C.

Advertisements.

MORTON'S
HANDBOOKS OF THE FARM.

The aim of the Series is to display the means best calculated to secure an intelligent development of the resources of our soil, and with the assistance which advanced Chemical investigation provides, to direct those engaged in Agricultural Industry towards the most successful results. Each Book is complete in itself, and the short Series of handy volumes, by various writers, who have been specially selected, forms a complete HANDBOOK OF THE FARM, which is abreast of the enterprising man's every-day requirements, and enables him economically to utilise the advantages which an everwidening science places within his reach.

PRICE 2s. 6d. EACH.

No. I. CHEMISTRY OF THE FARM.
By R. WARINGTON, F.R.S.
Revised and Enlarged. Eleventh Edition.

No. II. LIVE STOCK.
By W. T. CARRINGTON, G. GILBERT, J. C. MORTON, GILBERT MURRAY, SANDERS SPENCER, and J. WORTLEY-AXE.

No. III. THE CROPS.
By T. BORWICK, J. BUCKMAN, W. T. CARRINGTON, J. C. MORTON, G. MURRAY, J. SCOTT, and R. HENRY REW.

No. IV. THE SOIL.
By Professor SCOTT and J. C. MORTON.

No. V. PLANT LIFE.
By MAXWELL T. MASTERS, F.R.S.

No. VI. EQUIPMENT.
By WM. BURNESS, J. C. MORTON, and GILBERT MURRAY.

No. VII. THE DAIRY.
By JAMES LONG and J. C. MORTON.
Revised and Enlarged.

No. VIII. ANIMAL LIFE.
By Professor BROWN, C.B.

No. IX. LABOUR.
By J. C. MORTON.

No. X. WORKMAN'S TECHNICAL INSTRUCTOR.
By WALTER J. MALDEN.
Illustrated.

In crown 8vo volumes, the complete set of ten volumes, if ordered direct from the Office, carriage free for £1 2s. 6d.

VINTON & CO., LTD., 9, New Bridge Street, London, E.C.

www.ingramcontent.com/pod-product-compliance
Lightning Source LLC
Chambersburg PA
CBHW081717220526
45468CB00008B/1878